Science, Consciousness
and Ultimate Reality

Science,
Consciousness
and Ultimate Reality

edited by

David Lorimer

ia

Published in the UK by Imprint Academic
PO Box 200, Exeter EX5 5YX, UK

Published in the USA by Imprint Academic
Philosophy Documentation Center
PO Box 7147, Charlottesville, VA 22906-7147, USA

ISBN 0 907845 79 7

A CIP catalogue record for this book is available from the
British Library and US Library of Congress

Cover Illustration:
Fire Phoenix 2000 (detail)
Batik Painting by Thetis Blacker ©

TABLE OF CONTENTS

CONTRIBUTORS

Denis Alexander was an open scholar at Oxford University where he read biochemistry before carrying out research for a PhD in neurochemistry at the Institute of Psychiatry, University of London. Following this he spent 15 years in academic positions in the Middle East, latterly (1981-86) as Associate Professor of Biochemistry at the American University of Beirut, Lebanon. Upon his return to the UK he worked at the Imperial Cancer Research Fund, London, and since 1989 at The Babraham Institute, Cambridge, where he is currently Chairman of the Molecular Immunology Programme. He has published numerous articles and reviews, particularly in the research field of lymphocyte signalling and development. He is a Fellow of St. Edmund's College, Cambridge and editor of the journal *Science & Christian Belief*. His first book on science and faith was *Beyond Science* (1972). More recently he has written the critically acclaimed book *Rebuilding the Matrix - Science and Faith in the 21st Century* (2001) which provides a general overview of the science-religion debate. Denis Alexander lectures and broadcasts widely on the subject of science and religion.

Bernard Carr was a student at Trinity College, Cambridge, and wrote his PhD on the first second of the universe. He was elected to a Fellowship at Trinity in 1976, after which he spent a year travelling around America as a Lindemann Fellow before taking up a Senior Research Fellowship in 1980 at the Institute of Astronomy in Cambridge. In 1985 he moved to the Queen Mary College, University of London, where he is now Professor of Mathematics and Astronomy. He has also held Visiting Professorships at various institutes around the world. His professional area of research is cosmology, with particular interest in such topics as the early universe, dark matter and the anthropic principle. He also has a long-standing interest in the interface between science and religion, currently being the coholder of a grant from the Templeton Foundation for a project entitled 'Fundamental Physics, Cosmology and the Problem of our Existence'. For many years he has been an active member of the Society for Psychical Research and he is currently its President. He is the author of around 200 scientific papers on various aspects of cosmology and the anthropic principle, including the

pioneering *Nature* article, 'The Anthropic Principle and the
Structure of the Physical World' (1979), which he coauthored
with Martin Rees. He is currently editing a book on the subject
for Cambridge University Press, *Universe or Multiverse?* His
books in this area include *Mankind Humbled in the Progress of Science* (1985), *The Uroborus of Science* (1989), *Science and the Divine*
(1990) and *Life in the Universe* (1993). His papers on psychical
research include 'Can Physics be Extended to Accomodate Psi?'
(2001) and his Presidential Address to the Society for Psychical
Research 'Worlds Apart: Can Psychical Research Bridge the Gulf
between Matter and Mind?' (2002).

Chris Clarke studied at Cambridge, from 1963 to 1970, obtaining
BA (1st), Part III and PhD in Global Aspects of General Relativity; he then went on to hold a research fellowship there until
1974, when he took up a lectureship at the University of York. He
was Professor of Applied Mathematics from 1986 to 1999, and
Dean of the Faculty of Mathematical Studies for three years, at
the University of Southampton, where he is now visiting Professor. He has been member of the Editorial Boards of *J. Phys. A* and
Classical and Quantum Gravity, as well as deputy editor of the latter. He has served on the York Diocesan Synod of the Church of
England has been chair of GreenSpirit (the Association for Creation Spirituality). He was Chair of the Scientific and Medical
Network from 2000-2002. He is author of *Elementary General Relativity*, (1979), *Relativity on Curved Manifolds* (1990, with F de
Felice), *The Analysis of Space-Time Singularities* (1994), *Reality
Through the Looking Glass* (1995) and *Living in Connection* (2003).

Guy Claxton is Visiting Professor of Learning Sciences at the
University of Bristol Graduate School of Education. A pioneer of
'contemplative' or 'transpersonal psychology' in the UK, his
books include *Wholly Human: Eastern and Western Visions of the
Self and Its Perfection* (1981), *Beyond Therapy* (1986), *Noises from the
Darkroom: The Science and Mystery of the Mind* (1994), and *The Psychology of Awakening* (with Stephen Batchelor and Gay Watson,
2001). Seeing the foundations of 'proximal spirituality' in an
enthusiastic and meticulous attitude towards inquiry has led
him to become a leader of the 'learning to learn' movement in
schools, in which guise he currently lectures around the world.
Consultant on creativity and intuition to many companies and

business schools, Guy Claxton took a double first in natural sciences from Cambridge, and holds a doctorate in experimental psychology from Oxford.

Peter Fenwick was educated at Trinity College, Cambridge where he obtained an Honours Degree in Natural Science. His clinical medical training was carried out at St. Thomas's Hospital in London. After obtaining experience in neuro-surgery he specialised in psychiatry. He was Senior Lecturer at the Institute of Psychiatry, Consultant Neurophysiologist at Radcliffe Infirmary in Oxford, and Honorary Consultant in Neurophysiology to Broadmoor Special Hospital. He has published numerous scientific papers on brain function and also several papers on meditation and altered states of consciousness. He is President of the Scientific and Medical Network and also of the U.K. branch of the International Association of Near-Death Studies (The Horizon Research Foundation), reflecting his special interest in this field. He lectures widely in England, on the Continent and in the United States on brain disorders and has made many appearances on radio and television. He has published over 200 articles in scientific journals. He has written a series of books with his wife Elizabeth: *The Truth in the Light* (1994), *The Hidden Door* (1996), *Past Life Memories* (1999).

David Fontana is Visiting Professor of Transpersonal Psychology at Liverpool John Moores University and President of the Transpersonal Section of the British Psychological Society. For many years he has studied the relationship between Western and Eastern psychological systems, together with methods for deepening and expanding consciousness, and has written widely on dreams, meditation and psycho-spirituality. His most recent book is *Psychology, Religion and Spirituality* (Blackwell, 2003). He is the author of over 20 books on various aspects of psychology (including personal development) which together have been translated into 25 languages. Among them are *Psychology for Teachers, Social Skills at Work, Managing Stress, Your Growing Child, The Lotus in the City, Growing Together*. He has written over a hundred articles in journals. He is currently writing a major book on death and survival.

John Habgood did postgraduate research and teaching in physiology in the University of Cambridge from 1948-53. He was ordained priest in the Church of England in 1955. He was Principal of Queen's College, Birmingham from 1967-73 and Bishop of Durham from 1973-83. He was then Archbishop of York from 1983-95 and was created a life peer in 1995. He was also Chairman of the UK Xenotransplantation Interim Regulatory Authority. John Habgood is the author of nine books on science, religion, ethics and public affairs. His *Varieties of Unbelief* formed the Bampton Lectures in the University of Oxford. His 1997 Riddell Lectures in the University of Newcastle were published under the title of *Being a Person*. He was Gifford Lecturer in the University of Aberdeen on *The Concept of Nature*, the book of which was published in 2002.

David Lorimer is Programme Director of the Scientific and Medical Network, with whom he has worked since 1986. Originally a merchant banker then a teacher of philosophy and modern languages at Winchester College, he is the author of *Survival? Body, Mind and Death in the Light of Psychic Experience* (1984), and *Whole in One: The Near-Death Experience and the Ethic of Interconnectedness* (1990). He is editor of *The Spirit of Science* (1998), *Wider Horizons* (1998), and *Thinking beyond the Brain* (2001). He is Vice-President of the Swedenborg Society and the Horizon Foundation (The International Association for Near-Death Studies UK). He is Chair of Wrekin Trust and of the All Hallows House Foundation. He has a long-standing interest in perennial wisdom and has translated and edited books about the Bulgarian sage Peter Deunov. He is a Fellow of the International Forum and his most recent book – *Radical Prince: The Practical Vision of the Prince of Wales* was published in 2003.

Mary Midgley is a professional philosopher whose special interests are in the relations of humans to the rest of nature (particularly in the status of animals), in the sources of morality, and in the relation between science and religion (particularly in cases where science becomes a religion). Until retirement she was a Senior Lecturer in Philosophy at the University of Newcastle on Tyne, where she still lives. Her best known publications are *Beast and Man, Evolution as a Religion, Science as Salvation, Wickedness,* and *Utopias, Dolphins and Computers, Science and Poetry.* Her most recent book is *Myths We Live By.*

Ravi Ravindra obtained his B.Sc.(1959) in Geology & Geophysics, and Master of Technology (1961) in Exploration Geophysics from the Indian Institute of Technology, Kharagpur. He then gained a Master of Science (1962) and Ph.D. (1965) in Physics from the University of Toronto; and a Master of Arts (1968) in Philosophy from Dalhousie University, Halifax. He held post-doctoral Fellowships in Physics (University of Toronto, 1965), History and Philosophy of Science (Princeton University 1968-69) and Religion (Columbia University 1973-74). From 1979-2003 he was Professor of Comparative Religion and Adjunct Professor of Physics, Dalhousie University and since 1988 Professor of International Development Studies, Dalhousie. He is author of *Krishnamurti in the Long Line of Rishis in India* (2002), *Science and the Sacred* (2000), *Heart Without Measure: Gurdjieff Work with Madame Jeanne de Salzmann* (1999), *Yoga and the Teaching of Krishna* (1998), *Krishnamurti: Two Birds on One Tree* (1995), (Editor and Principal Author): *Science and Spirit* (1991) {Fourteen out of twenty-five papers are by R. Ravindra}, *The Yoga of the Christ in the Gospel According to St. John* (1990), *Whispers from the Other Shore: Spiritual Search East and West* (1984). His most recent publication is *Centered Self without Being Self-Centered.*

Alan Torrance is Professor of Systematic Theology in the University of St. Andrews. He was a lecturer in the University of Aberdeen from 1983-87, then held the Chair of Systematic Theology in the University of Otago from 1987-93. He then lectured at King's College, London before moving to St. Andrews in 1999. He was ordained into the ministry of Word and Sacrament in the Church of Scotland in 1984 and has served on a substantial number of church and ecumenical committees. He has also served on a number of editorial boards. He is author of *Persons in Communion: an Essay on Trinitarian Description and Human Participation* (1996), ed. (with John England) *Doing Theology with Stories of the Spirit's Movement in Asia,* (1991), trans. (with Bruce Hamill), with introductory essay, Eberhard Jüngel, *Christ, Justice and Peace* (from German: *Mit Frieden Staat zu Machen*), (1992), ed. (with Hilary Regan), with introductory essay *Christ and Context* (1993).

Forthcoming: *The Christ of History and the Open Society: the 1998 Hensley Henson Lectures, University of Oxford.* Edited books forthcoming: *God and Science, Ethics and the Doctrine of God, Theology and Gender.*

Keith Ward was Regius Professor of Divinity in the University of Oxford from 1991–2003. He is Emeritus Student of Christ Church, Oxford. He is an Honorary Fellow of Trinity Hall, Cambridge, and an Honorary Fellow of the University of Wales, University College, Cardiff. He is a Member of Governing Council, Royal Institute of Philosophy and a Member of the Editorial Boards of *Religious Studies; Journal of Contemporary Religion; Studies in Inter-Religious Dialogue; World Faiths Encounter; Teaching Theology and Religion.* He gave the Gifford Lectures in the University of Glasgow, 1993-1994. His main publications are: *Ethics and Christianity* (1970), *Kant's View of Ethics* (1972), *The Concept of God* (1974), *The Divine Image* (1976), *The Christian Way* (1976), *Holding Fast to God* (1982), *Rational Theology and the Creativity of God* (1984), *The Living God* (1984), *The Turn of the Tide* (1986), *The Rule of Love* (1989), *Divine Action* (1990), *A Vision to Pursue* (1991), *Defending the Soul* (1993), *Concepts of God* (1993). His four-volume 'comparative theology' was published between 1996 and 2000: *Religion and Creation, Religion and Human Nature, Religion and Revelation* and *Religion and Community.* Other more recent publications are: *God, Chance and Necessity* (1996), *God, Faith and the New Millennium* (1998) and *A Short Introduction to Christianity* (2000), *God: A Guide for the Perplexed* (2002) and *The Case for Religion* (2004).

David Lorimer

Editor's Introduction

A parallel principle drives both science and mysticism — the assumption that unity lies at the heart of our world and that it can be discovered and experienced by man.

Renée Weber

We are within a reality that is also within us.

Plotinus

The Science, Consciousness & Ultimate Reality Project

This volume arises out of the events held by the Scientific and Medical Network between November 2001 and July 2003. The aim of the 'Science, Consciousness and Ultimate Reality' project, supported by a generous grant from the John Templeton Foundation, was to examine issues at the interface between science, religion and the relatively new field of 'consciousness studies'. When I sent the series booklet to Anglican bishops, one replied that the title gave scope for almost any range of topics! However, all the events were related to this rubric and titles included *Mind, Brain and Beyond; Science and Human Values; Consciousness and Human Responsibility; Spirituality, Science and Religious Experience; Religion, Science and a New World-View;* and *The Mind and its Place in Nature.* The project was arranged into five phases:

1. A series of public dialogues at twelve different UK universities. The universities were Oxford, Cambridge, King's Col-

lege London, Bristol, Kent, Trinity College, Dublin, Wales (Lampeter), Durham, York, Edinburgh, Glasgow and St. Andrews. Speakers and chairs at these events, in addition to those featured in this volume, included Lord Sutherland of Houndwood FBA, Prof. Chris Isham, Prof. Max Velmans, Prof. Stephen Clark, Prof. Richard Gregory FRS, Sir John Polkinghorne FRS, Prof. John Hedley Brooke, Lord Robert Winston and Sir Brian Heap FRS.

2. Two seminars at Durham University for thirty-five students, held in July 2002 and 2003.

3. An invited seminar in June 2002. This furthered the issues raised in the dialogues and served as a preparation for the public event to follow.

4. A two-day public conference at King's College, London in June 2002.

5. A CD-Rom for distribution to universities, schools and interested institutions (may be ordered via www.scimednet.org and contains full reports of all meetings plus an extensive resource section).

The emphasis throughout the project was to stimulate serious interest in the Science, Religion and Consciousness field. It is important to move beyond the stereotypes of science vs. religion so prevalent in the media and to extend the science–religion debate to include consideration of the nature of spiritual experience and of interdisciplinary issues arising from consciousness studies.

Science and Consciousness

'What is consciousness?' or 'Who is conscious?' are perhaps two of the most fundamental questions that we can ask. We can all try to answer these questions from our own experience, but what do science, psychology and religion have to say? The 'Science, Consciousness and Ultimate Reality' project was designed to explore these questions in a systematic way.

The 1990s were declared the Decade of the Brain and saw a very significant increase in scientific interest in the study of consciousness after decades of neglect owing partly to the pervasive influence of behaviourism in academic psychology. Major international conferences on science and consciousness have been held around the world, notably those based at the University of Arizona in Tucson which have attracted hundreds of delegates. The *Journal of Consciousness Studies* was founded in 1994 and is now well established ten years on.

Some of the vital pioneering work in the analysis of mystical and spiritual experiences was carried out by William James a hundred years ago and written up in his classic book *The Varieties of Religious Experience.* This work was later followed up in the 70s by the biologist Sir Alister Hardy, who received the 1985 Templeton Prize for his work summed up in his book *The Spiritual Nature of Man.* The study of religious and spiritual experience has in turn become part of the field of transpersonal psychology. Such studies postulate the existence — or at any rate the experience of sacred or numinous realms that seem to transcend the physical dimension and normal states of human awareness. These studies have been influential in expanding contemporary views of the mind and consciousness.

The American philosopher Ken Wilber has summarised the various schools or approaches to consciousness under twelve headings.

- Cognitive Science
- Introspectionism
- Neuropsychology
- Individual Psychotherapy
- Social Psychology
- Clinical Psychiatry
- Developmental Psychology
- Psychosomatic Medicine
- Nonordinary States of Consciousness
- Eastern and Contemplative Traditions
- Quantum Consciousness
- Subtle Energies Research

This list conveys something of the breadth of the field, ranging from the 'hard' to the 'soft' end of the spectrum. Each area has its own literature and makes its own contribution towards developing what he calls an integral view. Some areas have a longer academic pedigree than others but this is partly due to the relative underdevelopment of subjectively based approaches.

Among those currently involved in consciousness studies there are two basic orientations: one follows the traditional Western method of looking from the outside in as detached observers — the third-person perspective. On the other hand, there are those who 'look from the inside out' — the first-person perspective — and who are interested in exploring the nature of their own consciousness. Eastern meditative traditions fall into this latter category, and, as Ravi Ravindra explains in this volume, yoga has its own form of objectivity through rigorous self-observation and training of the mind. The third-person group focuses on experiment while the first-person approach is more concerned with direct experience.

Another radical distinction of approach, not unrelated to the contrast between first- and third-person perspectives, can be drawn between those who assume that consciousness is entirely dependent on brain processes and those who contend that consciousness may in some sense be 'beyond the brain'. In his Ingersoll Lecture on Immortality that draws upon the ideas of the Oxford philosopher F.C.S. Schiller, William James contrasts theories whereby consciousness is thought actually to be produced by brain processes with those positing that consciousness is in some sense filtered by or transmitted through the brain. The former 'productive' theories lead inexorably to the prediction that consciousness is extinguished at bodily death, while the latter 'transmissive' theories leave open the possibility that an aspect of the self may survive the death of the brain. 'Productive' theories necessarily regard brain processes as causing conscious experiences, while 'transmissive' theorists insist that this apparent causation may in fact represent no more than a correlation between brain processes and conscious subjective experience.

Why are consciousness studies important? Because they mark a significant stage in the growing self-awareness of science, a science that can no longer takes its philosophical presuppositions for granted. Comparative religion, anthropology, sociology, post-modernism and the work on scientific revolutions by Thomas Kuhn and others have all contributed to the interdisciplinary understanding that modern science is underpinned by a series of interconnected and rarely acknowledged assumptions that impact on the very definition of the human being.

Among the issues addressed during the project were the following:

- How might a new science of consciousness contribute to moral, spiritual and scientific progress?

- Is there any validity in means of obtaining knowledge other than through the five senses?

- What is the relationship between the self or consciousness and the brain? How do you justify your position on this question and on what basis do you reject other positions? Is there any evidence-value in near-death experiences that might relate to this question?

- Which data from mystical experiences might challenge current materialistic assumptions and which of these data might validate propositions about the existence of God and a transcendent dimension as the source of values?

- How might one account for the transformative effects of mystical and near-death experiences?

- What evidence is there for purpose and meaning in the universe?

- What might be recognised as a point of spiritual convergence between the great religions — or possibly even beyond them, including science, thus giving a new perspective on science?

- On what basis do you derive your own deepest values and sense of meaning?

- What role do philosophical assumptions play in science and religion?

- Is epistemological and ontological reductionism a necessary assumption in science or a hypothesis to be justified and applied to appropriate fields?

A number of these themes reappear in the papers below, as we shall see.

Structure of the Volume

> It is better to debate a question without settling it than to set-
> tle a question without debating it.
>
> *Joseph Joubert, 18th century essayist*

The volume falls into three parts: physicsand consciousness,
neuroscience and psychology, and theology and ethics.

The first section is enriched by the contrast between the eastern and western approaches of its authors, all of whom are
informed by their own distinctive understandings of spirituality
in relation to science. It seems fitting to begin on the largest scale
— that of mind and the cosmos, discussed by Bernard Carr. As
an astrophysicist who is also the current President of the Society
for Psychical Research, he is especially well placed to link the
two. His essay asks the question: is mind fundamental or incidental, a question central to this volume and which is for the
most part sympathetic to the former view. Carr's essay falls into
three parts. Part 1 describes the progress of cosmology involving
a progressive realisation of the immensity of the physical universe and a corresponding displacement of humans from the
centre to a remote periphery. The vaster the universe (understood in mechanistic terms) the more insignificant does humanity appear. Quantum theory, discussed in more detail by Chris
Clarke below, shattered our atomistic theories of the microworld.

At this point Carr introduces his own original contribution in
the form of his Uroborus diagram, a symbol of the snake biting
its own tail in which he maps the scales of the very large onto the
very small, showing the extent of interconnectedness of levels. It
also leads into his second part, discussing the scientific evidence
for the significance of mind and arguing against the conclusions
of the first part. Human beings occupy an interesting position on
the Uroborus scale. The limits of our perception have been constantly expanding with the development of ever more sophisticated telescopes and microscopes. This leads Carr into a
consideration of the various forms of the Anthropic Principle,
which remains a contentious issue within science but which is
consistent with the recognised complexification of mind in the
course of evolution. He then sets out his view of the arguments
for the significance of mind in the context of paradigm shifts that
have occurred in physics since Newton:

- Man — or at least complexity — is central
- Life is fundamental to the Universe
- The blossoming of mind is fundamental to the cosmos
- The unity and beauty of the Universe point to a guiding intelligence

Carr's last avenue of exploration is scientifically the most controversial, namely psychical research, which dates back in a formal sense to the founding of the Society for Psychical Research in 1882 by Fellows of Trinity College Cambridge, of which Carr is also a Fellow. The thrust of science at that time was towards an all-embracing naturalism, a world-view still held by many scientists but which makes the incorporation of consciousness problematic. The vast majority of those who have closely examined the data from psychical research have concluded that there is a real effect to explain, but there are bodies such as CSICOP (the committee for the scientific investigation of claims of the paranormal) who still like to claim that there is no such thing as psi. Carr indicates various areas where there is considerable evidence of the anomalous interaction of consciousness with the physical world, remarking that such phenomena cannot be accounted for by tinkering with the current paradigm, but that physics must be explicitly extended to incorporate consciousness.

Carr therefore proposes that an adequate new explanation 'must transcend our usual concepts of space and time and involve some form of higher dimensional reality structure ... because, as we now show, many psychic phenomena seem to involve a communal space, which is not the same as ordinary physical space but subtly interacts with it.' This means that physical and mental space must in some way be combined since psi indicates a blurring or overlap of these spaces. Carr puts forward the view that percepts exist in Kaluza-Klein space and that 'the physical Universe is as 4-dimensional "brane" embedded in a higher dimensional "bulk".' Logically, extra dimensions of space are required in order to account for psi phenomena, so it seems to me that Carr is theorising along very promising lines, even if he is leaving his colleagues behind.

Chris Clarke then introduces one of the leitmotifs of the book, namely the limitations of a dualistic interpretation of reality, especially in relation to the mind–body problem. Readers will find that a number of irreconcilable views on this problem are set out in this volume, but in ways that will nevertheless challenge their thinking. Clarke shows how atomistic and mechanistic metaphors are closely allied with dualism and describes the demise of atomism, defined as 'the idea that we need to look to the small scale level of particles in order to find reality.' He asserts that there is no ultimate reality to be found at the microscopic level, maintaining that quantum non-locality thoroughly undermines atomism. This leads him to the stronger position of maintaining that 'physics cannot provide any sort of ultimate reality at all' although it can specify a direction for a more complete understanding of reality.

Besides atomism, the other notion investigated — and redefined — by Clarke is that of (wave–particle) complementarity, a word that he feels should be replaced by 'incompatibility'. Clarke contends that physics itself is incomplete unless one introduces consciousness, but in what rigorous sense can it be introduced? He points out that many such approaches are underpinned by 'interactionist dualism, where consciousness is regarded as a separate layer of reality from the physical'. He then asks: 'what exactly is it that consciousness is supposed to do, that material entities cannot do, and how exactly does an immaterial consciousness interact with a material universe?' His own approach follows the distinction between first- and third-person theories already mentioned and spelt out by Max Velmans. Third-person approaches in fact presuppose first person experience:

> If consciousness is taken as a primitive element in the theory (and all theories must have some primitive elements) then we can define third-person, objective concepts in terms of first-person concepts by extracting those elements of first-person views that people share in common. By reversing the usual order of argument, a way is opened up (and one that is not fundamentally flawed at the outset) for determining which things are conscious.

Consciousness fits into physics as a means of 'establishing the patterns of meaning within which physical reality manifests itself; and quantum physics presents precisely the appropriate amount of indeterminism and flexibility to allow consciousness to do this.' Instead of mind-matter interaction, Clarke suggests that that the distinction between these aspects is not so much functional as based on first-person meaning and third-person dynamics, a view that in some ways reflects David Bohm's notion of soma-significance. We are co-creators involved in a participatory universe and therefore intrinsically connected to each other. This view overcomes alienation and restores meaning: we are separated neither from the world nor from each other, as Mary Midgley also observes. Clarke concludes by saying that such a picture affirms rather than denies our humanity while being consistent with existing science in a way that combines the necessary ethical values with scientific knowledge.

Ravi Ravindra takes epistemology as his starting point, sharpening a distinction of approach that is already implicit in Chris Clarke's paper. Science and yoga are both about 'objective knowledge', but while science takes its mental/theoretical constructs to be real, the yogi tries to cleanse his perceptions and remove such constructs in order to experience reality more directly. Needless to say, scientifically (or post-modern) minded commentators deny any purported experience of pure consciousness, insisting that all experience is by definition mediated and therefore (socially) constructed. Instead of seeking certain truth, science now theorises about probable truths.

Here we meet one of Ravindra's essential points: that quantitative extension of perception through experiment is not equivalent to its qualitative transformation. The knower is not transformed from the 'rational cogniser' of modern science and philosophy but remains resolutely within his existing categories. Here Ravindra makes a point taken up at greater length by Mary Midgley, namely that feelings and sensations are downgraded as means of acquiring knowledge. He concludes: 'the task of yoga, and of all spiritual disciplines, is not the same as the task of the scientific inquiry. Whereas science seeks to understand and control processes in the world, using the rational mind as the tool of exploration and explanation, yoga seeks to transform the human being so that the reality behind the world can be experienced.' Ravindra then explains this tradition in more detail.

In his discussion of the body and the embodied, Ravindra describes the progressive change of perception of the self as the body towards the 'self as inhabiting the body', which means that death is 'giving up the body'. Here Ravindra appears to part company with Clarke, Midgley and Claxton (but aligns himself

with Fenwick) by supposing that the self is in some sense detachable from the physical body. However he elaborates the meaning
of *sharira*, normally translated as body, by pointing out that 'it
means the whole psychosomatic complex of the body, mind, and
heart. *Sharira* is both the instrument of transformation as well as
the mirror indicating it.' This position is therefore altogether
more subtle.

Ravindra asserts that the whole purpose of yoga is to align the
body–mind more precisely to the purposes of Spirit. The mind
(the rational cogniser) is not the true knower but 'is confined to
the modes of judgement, comparison, discursive knowledge,
association, imagination, dreaming, and memory through which
it clings to the past and future dimensions of time. The mind
with these functions and qualities is limited in scope and cannot
know the objective truth about anything.' Hence the discursive
mind needs to be stilled so that undistorted perception — real
seeing — may take place. This highlights a parallel distinction
between the mind (*chitta*) and the real Seer (*purusha*). When the
mind is emptied of itself, then *samadhi* can arise — 'a state of
silence, settled intelligence, and emptied mind, in which the
mind becomes the object to which it attends, and reflects it truly,
as it is.' And with this comes peace, cessation of agitation. The
deepest self cannot be known as such (the lower or more limited
cannot fully know the higher or more limitless), but it can be
identified with — in participatory knowledge by identity,
another octave of what Chris Clarke calls 'living in connection'.

Ravindra concludes that a new science of consciousness
requires transformed scientists and that there are tried and
tested methods for achieving such transformation: 'the subtler
and higher aspects of the cosmos can be understood only by the
subtler and higher levels within humans.' Hence refinement of
the human instrument is essential, as Ken Wilber and others
have also argued. As Ravindra puts it: 'the important lesson here
from the perspective of any future science of consciousness is the
importance of knowledge by identity. We cannot remain separate and detached if we wish to understand. We need to participate in and be one with what we wish to understand.' This
exercise implies both an overcoming of the intrinsic isolation of
the ego and a thoroughgoing epistemological reversal of three
centuries of outer-directed, object-based thinking while still
maintaining the analytical rigour gained in the process. More-

over, a higher, enlarged or more developed consciousness also implies the same for conscience — hence for values — and a corresponding expansion of the compassion and the heart: wisdom allied to love. In this way, Ravindra suggests, we may achieve a realisation of what he calls the 'First Person Universal'.

Neuroscience and Psychology

Peter Fenwick played a major role in the conference series, speaking on several occasions in different contexts. He is unusual within the neuroscience community in being exceptionally well informed about the nature of spiritual and near-death experiences. This has led him to take the view that we need a wholly new understanding of the dying process that does justice to the patterns emerging from empirical findings. It has also made him question the unstated premise of modern neuroscience, namely that 'psychological processes are generated entirely within the brain and limited to the brain and the organism.'

Fenwick maintains, rightly in my view, that the way forward for a science of consciousness must involve analysis of the metaphysical foundations of modern science, the consequence of which is that consciousness is excluded from the picture and can only 'added in as an extra which somehow arises.' He goes on: 'but as the whole basis of the perception of life and our understanding of the world is within consciousness, it is highly unlikely that there will be a formula which will allow it to arise unless consciousness in some way is given a part in our theories.' Here Fenwick is reinforcing a central message of this volume: that we need a radical reconfiguring of our understanding of mind and consciousness that does justice to subjective experience. Even the most refined neuroimaging techniques can only display *neural correlates,* but correlates are not equivalent to causes.

Fenwick spells out the requirements of this new approach:

> In my view, this new science of consciousness must include a detailed role for brain mechanisms, an explanation for the action of mind outside the brain, and an explanation of free will, meaning and purpose. It should also give an explanation of wide mental states, including mystical experience and near-death experiences, when the experiencer sees through

into the structure of the universe. Finally, it should provide a clear explanation for apparent downward causation (purpose) throughout the universe and in the brain, as well as some solution to the questions raised, particularly in Eastern cultures, of the survival of aspects of consciousness after death.

After making a connection with Chris Clarke's view and commenting on the implications of Amit Goswami's idealistic universal consciousness approach, he spends the rest of his paper addressing his own requirements, beginning with mystical experience and moving on to prayer studies. Light is one of the recurring themes in mystical experience. It is usually brilliant white or gold, is felt as embracing and warm, and is experienced concurrently as love. Love is intertwined with light in the experience and is experienced as universal, interpenetrating everything. Fenwick takes the view (not one shared by Guy Claxton, as we shall see) that the sense of reality associated with these experiences is an (ontological) indication of a dimension of conscious existence beyond the physical. These experiences cannot simply be reduced to localised neuronal firing.

With respect to where consciousness might be located in the brain, Fenwick asked in his lectures: was it in the cells? No. Was it to be found in groups of cells? No. Was it in a part of the brain? No, because its manifestation is in fact widely distributed rather than precise. Perhaps consciousness can be located into dorsolateral prefrontal cortex associated with working memory? This is an area that is affected (through low blood perfusion) in patients with dementia who have no sense of self. Again the answer is no.

Fenwick calls the modular brain identity theory 'the new phrenology'. It proposes that the brain itself is modular, that each function of the brain is modular and that the mind (in so far as it is defined as such) is also modular. There are, however, major difficulties with this view:

- It has no explanation of modularity
- It has no explanation of subjectivity
- It has no explanation of consciousness
- It gives no indication of how to move from neural firing to experience

This is a serious list of shortcomings and one that corresponds to the admission in Ian Glyn's *Anatomy of Thought* that the connection between brain and subjective experience is 'una granda lacuna'. Correlations can be established but these do not logically amount to causes, as we have already observed.

This leads Fenwick on to a consideration of the dying process, which he has studied for many years (he is President of the Horizon Foundation, formerly the International Association for Near-Death Studies UK). He begins by observing that 'one of the central planks of scientific materialism is that when the brain dies, consciousness ends'. He divides his analysis into the phases of the Nearing Death Experience and the Near Death Experience (NDE). In the first category he gives examples of 'take away visions' in which patients report visions of people (often friends or relatives) who ostensibly appear to take them over the threshold. Some seem to be living in two worlds at once shortly before death. Death bed coincidences — where relatives see an apparition coinciding with death — also occur and were well documented as long ago as 1886 in the two-volume study *Phantasms of the Living* by Gurney, Podmore and Myers.

Many people reporting visions experiences become (or are already) lucid at the time of the experience and die immediately afterwards. Osis and Haraldsson found that 13% of take away visions were of religious figures, 17% were alive while the vast majority — 70% — were apparitions of the dead. Other studies indicate that 30% of those apparently seeing deathbed visions die getting out of bed as if they were trying to go somewhere. Fenwick gives a number of specific examples of the phenomenon and also draws on some Italian research where 40% of deathbed visions were of the take away variety. Preliminary surveys at hospices indicate that these visions are far more frequent than commonly supposed, but more research is required.

Fenwick proposes that the best model of near-death experience (NDE) is that of cardiac arrest, following his own research and that of the Dutch cardiologist Pim van Lommel published in the Lancet in 2001. As he observes:

> For the scientific researcher, the interesting question is this: when does the NDE occur? Does it occur before or during unconsciousness caused by the heart attack, during recovery or after recovery? The onset of unconsciousness is very quick, as occurs in a faint, so the experience could not occur

then. During unconsciousness all brain functions cease so neuroscience says the experience could not occur then either. During recovery from a cardiac arrest the subject is confused so the clear, lucid NDE could not occur in this confusional state. If it could be shown scientifically that the near death experience occurs during unconsciousness, as suggested by those who have survived a cardiac arrest, when all brain function has ceased and there is apparently no mechanism to mediate it, this would be highly significant, because it would suggest that consciousness can indeed exist independently of a functioning brain.

There is indeed suggestive evidence that the NDE does occur during the period of cardiac arrest, but follow-up scientific studies now need to be carried out. Fenwick believes that 'science itself suggests that we may now need a different way of looking at the brain and at consciousness — a science of spirituality'. And he concludes: 'both transcendent experiences and the near death experience appear to give a subjective view of what lies beyond the physical, suggesting that the very structure of the world is spiritual, that consciousness is primary and unitary and that individual consciousness is part of the whole and survives death.'

This is not the way that Guy Claxton sees the same experiences:

Peter Fenwick and I rig the scales so that the same weight of evidence tips them in different ways. His apparent even-handedness masks his wish that the Out of the Body experiencer really does float up to the ceiling, and can thus see the secret message scrawled in the dust on the top of the high cupboard. I want it not to be so, and will ferret out the flaw in the experiment that allows my pretheoretical commitment to escape. Neither of us is innocent, and we would be better employed meditating on what we each need to be true, and why, than on trying to establish the impersonal truth, for we shall never agree on what is sufficient proof.

Claxton makes a crucial point here, which readers may wish to ponder: what are our own 'pretheoretical commitments'? How are these related to the metaphysical foundations of modern science? In what respects are we partial or selective in our assessment of evidence? Do we have reasons for wishing a particular view to be true — and why? Debates around these issues are dressed up in a number of ways: the limits of the 'scientific', the equation of 'scientific' with 'rational' and effectively 'materialistic', the judicious use of Occam's razor to exclude unwanted considerations (I call this questionable use Occam's hatchet); and spurious notions like assessment of the 'antecedent probability' of an event occurring, which is simply another way of restating pretheoretical commitments or metaphysical assumptions. On the other hand, we can cite a lack of rigour in assessing the validity of subjective evidence, the unreliability of witnesses, the fluctuation of consciousness and attention, the role of both the preconscious and the unconscious. All these factors are guaranteed to keep the debate alive, but we are each responsible for our own integrity and for maintaining a self-critical awareness of our interpretations of our own subjective experience and those of other people.

Claxton puts his case with great eloquence, arguing that 'the evidence from cognitive neuroscience further undermines any hubristic tendency to claim physical, independent reality for the things we experience, and demonstrates just how much of our perception is imputed rather than extracted.' We are not only embedded in culture, but subject to the vagaries of a limited instrument of perception: our perceptions are partial, reduced, selective and biased by our character and history.

This realisation makes Claxton cautious in drawing Fenwick's (or indeed my own) conclusions from mystical experience; he prefers not to reach for any supernatural interpretation, but see how far current neuroscientific models can be pushed. In what precise sense, he asks, is an experience 'real', and does this assertion make the subject's own interpretation more credible?

For Claxton, the central category error is the idea of a substantial self, 'the belief that behind their thinking lurks a thinker; behind their action an actor; behind their experience an experiencer, behind their decision a decider: that there is a ghost in their own machine more real and incontestable than the one in the mediaeval passageway.' Claxton eschews approaches to spirituality that involve grand metaphysics or the special status of visionary experiences, preferring a path that attempts to 'clean the gizmo through which [she] experiences the world, so that she can tell what are real stars, what are specks of dirt that have got stuck to the lens, and what are intrinsic limitations of the instrument — simply reflections of the kind of 'telescope' she happens to be.'

This is what Claxton means by the 'proximal spirituality' of his title, and here he joins Ravi Ravindra in his approach to 'cleansing the doors of perception' through attentional practices. These techniques have the effect of increasing the quality, intensity and immediacy of experience: 'proximal spirituality values these kinds of shift in the quality of experience, and is centrally concerned with contemplative methods that promote their development.' We cannot in the end 'rend the veil of illusion'. All we can aspire to is a 'relative cleansing' or removal of the sources of distortion, an awareness that our beliefs are pre-dissolved in experience. We may see more clearly but we are irredeemably bound to a partial view. This approach enables a deliberate move past the 'self-system' that may give a glimpse of what is there before our habits kick in. Moreover, as Claxton concludes, if living on the leading edge of experience was good enough for Jesus and the Buddha, then it is good enough for him.

David Fontana's paper brings us back towards Peter Fenwick's view of reality, but he does share Guy Claxton's concern about the 'extent to which our understanding of these facts is a product of our own patterns of thinking rather than of a direct encounter with something intrinsic to that universe itself', although he draws a different conclusion from this observation. After discussing the limitations of both scientific and psychological realism, Fontana goes on to make the case for transpersonal psychology, concentrating on the contribution of Ken Wilber.

He analyses Wilber's classification of inner- and outer-oriented groups, later elaborated into a model of four quadrants in which the inner and outer, individual and collective are represented in an all-encompassing scheme. Neither group has the full picture, so both aspects or orientations are required for a more comprehensive view. There is currently an imbalance, with much more attention focussed on the second group and its objective methodologies: 'we spend a great deal of time exploring the outer world revealed to us by our minds, and little or no time in exploring those minds themselves.'

Fontana goes on to a detailed consideration of two views on the mind-brain relationship. With respect to the outer-oriented group, He makes the point that 'the argument that we do not know how a non-physical mind can interact with a physical brain (clearly a major stumbling block for many scientists) does not invalidate the fact that it may do so — because we equally do

not know how the electro-chemical activity of the brain can produce non- physical events such as thoughts, intentions, volitions
and a moral sense.' In common with Peter Fenwick, he argues
that the evidence from parapsychology should make us take this
first question seriously, and he himself has taken part in extensive empirical investigations, notably the events leading to the
publication of the Scole Report by the Society for Psychical
Research — probably the best modern evidence for survival of
consciousness after death.

Theology and Ethics

The next section continues the debate about the mind–body
issue, starting with a lecture given by Keith Ward in Edinburgh.
Ward begins by saying that the concept of the soul had been popularly used to distinguish humans from animals before making a
preliminary distinction between Indian and Semitic usages of
the word. Indian philosophy drew a sharp distinction between
spirit and matter, arguing that there is one self (Ramana
Maharshi would say Self) that undergoes many embodiments or
incarnations. The true nature of the self is pure and in this view
the soul is a spiritual substance, a position also held by Plato
(perhaps following Indian influence, as argued by
Radhakrishnan in *Eastern Religions and Western Thought*).

Ward points out that the Semitic view of the soul is not originally substantial in this sense. The Biblical image is that of the
breath of life (*nephesh*), the active principle of the living body —
not a separate entity. This notion corresponds to Aristotelian use
of the soul as the form (*eidos*) of the living body. Aristotle's fundamental categories are therefore matter and form rather than
matter and spirit, with the soul as the blueprint or dynamic shaping principle. Aristotle distinguishes three levels of soul: vegetable, animal and rational or intellectual soul, all of which are
organising principles on different levels. The intellectual soul is
characterised by imagination issuing in thought (*theoria*) and free
responsible action within a social context. In modern parlance,
Aristotle is a dual-aspect monist, insisting that the soul cannot
exist apart from the body. Thus human beings can be understood
as 'material objects' with distinctive capacities. It follows that all
who have these capacities, whether exercised or not, partake in
human nature.

Ward shows how Aquinas takes up Aristotle's ideas and amends them to make them consistent with Christianity. Since Christians are not materialists, the possibility of a non-material or spiritual existence arises, all the more so since God or the Supreme Being is conceived of as spiritual and non-material — having form without matter. Moreover, the spirit assumes causal priority and entails consciousness and intention or will.

Augustine's scheme involves God attracting creation towards Himself through love so that creation tends towards Divine Perfection. An ambivalence arises with the concept of God as Spirit and humans made in His image, i.e. essentially spirit as well. This persuades Aquinas that the soul must be a 'substantive form', thus moving beyond Aristotle. However he is emphatic that 'my soul is not me'. We find Aristotle's scheme in Descartes, where, in the Sixth Meditation, he insists that the soul is a 'single whole' and not to be likened to a pilot on board a body. This is somewhat at odds with the radical dualism normally attributed to him, a point we shall find reappears in Mary Midgley's paper below.

Ward devotes some space to a discussion of bodies — physical and spiritual — and the resurrection. Much philosophical thought assumes that the only alternative to a physically embodied existence is a totally disembodied state, which is impossible to imagine and therefore considered nonsensical. However, every conceivable world of form presupposes the need for a corresponding body or else there is no possibility of manifestation let alone relationship. Paul's spiritual body (1 Corinthians 15) is germinated from the flesh but is incorruptible. Ward builds on this idea by suggesting that the soul or self is in a sense generated through a physical body but this must be transferred and indeed transformed for a post-mortem existence to be possible. This leads him to formulate the new position of 'emergent dualism' or qualified monism whereby the self/soul arises in conjunction with the physical body but is capable of surviving its dissolution. Here the soul is defined as a set of capacities developed over a lifetime.

Does the human soul have a special dignity? Ward answers in the affirmative. Whatever God has created has intrinsic value. We are integral parts of the natural world but we may qualify for immortality at a certain point in evolutionary development, we are 'called to immortality'. Immortality is an emergent property within a cosmos directed towards the Divine. This view does not confer Greek intrinsic immortality on the soul and it is consistent with a panpsychic view of evolution whereby consciousness is inherent in the universe and manifests as complexity unfolds. A number of problems remain with this view: at what point in development does one become capable of immortality? How is this squared with the traditional qualities of love and justice attributed to God? And is there any experiential evidence supporting this view?

Mary Midgley brings us back to the perennial mind–body problem and begins by asking: 'what does it mean to say that we have got a mind–body problem? Do we need to think of the relation between our inner and outer lives as business transacted between two separate items in this way, rather than between aspects of a whole person?' Her answer is no. Like Chris Clarke, she points out that seventeenth-century dualism assumes a view of physics that we no longer hold. She analyses some of the historical reasons for the mind-body split, including the divorce of reason from feeling and the need for the then new physics to differentiate itself from older ways of thinking.

She explains how the problem of consciousness was tackled respectively by the behaviourists (eliminating it), Colin McGinn (the mystery is irreducible), proposing that a better starting point might be the relationship between our inner and outer lives: 'between our subjective experience and the world that we know exists around us — in our experience as a whole, rather than trying to add consciousness as an afterthought to a physical world conceived on principles that don't leave room for it.' We have already encountered the problems of consciousness as an 'add-on' to the physical universe in Chris Clarke's and Peter Fenwick's papers.

Midgley devotes considerable space to understanding Descartes' view and its subsequent evolution through western philosophy, warning against prioritising immortality (and hence the separability of the soul from the physical body) in fashioning an integral account of ourselves. We need to dispense with the extreme abstractions implied by the very terms mind and body, and 'get right away from Descartes' idea that the inner life is essentially a simple thing, a unified, unchanging entity, an abstract point of consciousness.' She argues that the unity of the human being is in fact a highly complex and dynamic process involving the reconciliation of competing parts of ourselves.

In formulating her own approach to 'explaining' consciousness, Midgley reminds us of the wider context of organic life and movement within an unfolding evolutionary process, suggesting that consciousness is perhaps the superlative of life. The discontinuity lies in our own mode of functioning and the way in which we normally interact with humans — socially — as opposed to stones (Martin Buber's I–Thou as opposed to I–It). These modes have clear implications for values, attitudes and behaviour, as will become clearer in Denis Alexander's paper below. However, we should not be seduced by the lure of simplicity, whether from physics or biology, since it leads invariably to a reductionist explanation that is by definition incomplete and therefore unsatisfactory.

Midgley shows how the idea of the 'secluded soul' not only cuts us off from other people, but also from the earth itself, a move only now being reversed with the emergence of ideas like biological symbiosis and the inseparability of organism and environment proposed by the Gaia hypothesis, which sees the earth as a whole in terms of living systems. This serves as a reminder that consciousness studies cannot afford to ignore our biological and ecological contexts. We are neither separate nor self-sufficient, as Midgley puts it. And we would do well to recognise the parallels between the history of the mind-body problem and that of human's relationship to the earth.

Alan Torrance addresses the question: does naturalism in the cognitive sciences constitute a crisis for theism? Before coming to theism he gives an analysis of what he sees as a challenge, if not a crisis, for the academy in the form of two mutually incompatible (but nevertheless materialistic) approaches:

1. *Naturalism* — dominant in the natural sciences and defined by the propositions that, according to Roger Trigg, 'reality is wholly accessible (at least in principle) to the natural sciences. Nothing ... can exist beyond their reach'; and second (Alvin Plantinga) that 'there is no God, and we human beings are insignificant parts of a giant cosmic machine that proceeds in majestic indifference to us, our hopes and aspirations, our needs and desires, our sense of fairness, or fittingness'. This approach is typified by Dewey, Quine,

Davidson, Dawkins and Dennett among others, and has
already been discussed by Carr above. The determinism
implicit in this approach tends to play down agency and creativity. It also creates problems for Darwinians who insist the
evolutionary fitness is the ultimate yardstick of truth. What
is useful or beneficial may or may not be true in a more ultimate sense.

2. *Enlightenment humanism, creative anti-realism, social
   constructionism* — dominant in the arts and social sciences.
   With its origins in Kant, this approach maintains that human
   beings are fundamentally responsible for creating the structure and nature of the world — we are ultimately the architects of the rationality of the universe. Properties of objects
   are not intrinsic but rather creative human projections. Truth
   becomes 'the state of play' (Cupitt). The problem here, as
   Torrance indicates, is a crisis of truth, which seems to depend
   on the proposition that something is true if everyone believes
   it. The humanistic or constructive approach, in contrast with
   naturalism, overplays human creativity.

Torrance argues that both these approaches are seriously defi-
cient when compared with theism. Indeed they cannot even provide a mutually consistent account of each other's views. For
him theism provides an epistemic base of considerable explanatory power. It explains why there was something rather than
nothing, it enables access to truth, it provides an ontological
basis for values that can be embodied by human beings, and it
explains why the world is intelligible. This position, although
unfashionable these days, is nevertheless tenable. Unlike naturalism, it does at least provide a coherent framework for the
activities of the academy relating to the intelligibility of the natural order and the pursuit of truth.

Torrance then considers Jaegwon Kim's critique of non-
reductive physicalism whereby a mental event M is thought to
be responsible for a brain state P* which is its physical realisation. However the brain state P* is in fact caused by and antecedent brain state P, thus making M logically redundant. Mental
descriptions are then reduced to epiphenomena because of the
physiological processes, so we are back to the problems associated with add-on epiphenomenalism.

So can mental events supervene on causal interactions? Kim argues that no emergent causation (such as downward causation) can exist at all: all causality is explicable in terms of the causal relations between the most basic, sub-atomic components. Causality on this account does not and indeed cannot recognise individual entities of any kind, conscious or non-conscious. Given this view, it is not surprising that Kim concludes his 1998 *Philosophy of Mind* by saying that consciousness and mental causation are two 'intractable problems'.

Torrance sees a possible third alternative to physicalism and Cartesianism in Nancy Cartwright's pluralistic universe. Here patterns emerge in physical processes which have genuine causal powers. We are part of a highly complex universe characterised by a 'patchwork of laws'. This position repudiates 'nomological monism' (as in Kim), whereby there is only one type of law governing all events — in his case causal connections between basic particles. Cartwright — like Mary Midgley — also rejects the fundamentalist assumption that 'all facts must belong to one grand scheme', which means that the world cannot be explained in terms of the operation of a single kind of causal law. All this will make eminent sense to those familiar with systems theory with its notion of distinct but interlocking levels.

These considerations imply the fundamental question being addressed, namely 'where precisely are causal properties located?' Do they belong exclusively to physical objects? Perhaps the concept of a location to causal properties is itself a category error since causality is about relationships and the systems view implies a much more complex and distributed set of feedback loops than the original Cartesian idea of localised mechanical push.

Torrance concludes that academia is rent between the mutually incompatible fideisms of naturalism and creative anti-realism. He argues that theism can obviate problems associated with these views and that it sustains and justifies the academic search for truth. Physicalism in psychology cannot make sense of thought progression, and epiphenomenalism logically hoists psychologists on their own petard. Nomological monism needs to be repudiated since we live in a complex world that is not wholly amenable to reductive, naturalistic explanations. Cognitive science points to the need to reject naturalism and the scientistic fundamentalism of Richard Dawkins.

He argues that Christian Theism's explanatory power should not be underestimated. It recognises the complexity of selves as subjects not objects, subjects who are free, responsible personal agents who can reason and penetrate the intelligible structures of the world. Theism must also recognise physicality, that emotions can be chemically induced and that mental processes are subject to physical degeneration. Torrance does not claim to have arrived at a definitive account of agency and causality, but does show the direction in which a solution is likely to be found. He concludes that, if the agenda of a theistic account of human agency is challenging, then the challenges facing a naturalistic account are arguably greater still.

The last two papers take the debate beyond world-views to their impact on ethics and values. In a lecture given at St. Andrews under the overall title of Science and Human Values, Denis Alexander starts by remarking that contemporary western societies are profoundly ambivalent about science: on the one hand there is a vision of a high-tech universe in which we 'manipulate its powers to serve our own ends', and at the other extreme a vigorous anti- science lobby that regards scientists as dangerous meddlers hubristically playing God. Twenty-first-century science will certainly produce further dramatic advances that will throw up questions that science is poorly ill-equipped to address, especially when they concern notions of human identity and value. It is therefore an encouraging trend that significant amounts of scientific funding are now routinely available so that ethicists, philosophers and theologians can address pressing moral and ethical questions — for instance about our biological constitution — raised by these scientific advances.

In this context Alexander does not think it helpful that much of the tone of public debate is set by prominent hyper-reductionists such as James Watson, Peter Atkins and Richard Dawkins. Alexander quotes Watson as saying 'There are only atoms. Everything else is merely social work'. Such remarks go well beyond science and express a metaphysical scientism or materialism that is intrinsically hostile to human values, as is graphically expressed in John Gray's recent book *Straw Dogs* or as Alexander quotes from Dawkins: 'We are machines built by DNA whose purpose is to make more copies of the same DNA.... That is EXACTLY what we are for. We are machines for propagating

DNA, and the propagation of DNA is a self-sustaining process. It is every living object's sole reason for living'.

Talk of science and human values, Alexander argues, invariably brings up questions of the relationship between science and faith. He contends that humankind as a whole is 'incurably religious', and that models linking modernisation and secularisation had been found wanting. In order to tackle the question of science and human values, Alexander outlines four models of the relationship between science and faith before moving on to a consideration of their bearing on human values issues.

1. *The conflict model*
This view — dating back to the nineteenth century — is still popular with the media and the kind of hyper-reductionists already mentioned (and also some creationists in the US), but Alexander regards it is long since past its sell-by date, maintaining that no contemporary historian of science takes this model seriously. Nor is it historically true when applied to figures like Kepler, Newton, Boyle, Faraday and Clerk Maxwell.

2. *The NOMA model*
The second model derives from the work of Stephen Jay Gould on 'non-overlapping magisteria' whereby science addresses the empirical realm of fact and theory and religion extends to questions of ultimate meaning and moral value. Science and faith are effectively in separate watertight compartments and, since they are two different types of activity, there is no reason why they should come into conflict. The model does have its problems, notably the constant traffic of ideas between science and faith during the early scientific period; and believers within the scientific community do not on the whole wish to keep their science and faith apart.

3. *The integrationist model*
This view appeals more to working scientists since they can integrate scientific and religious knowledge. A good example of an integrationist thinker is Dr. Arthur Peacocke, who argues that a rational set of theological beliefs can be built on scientific knowledge and methods. A different view is held by E.O. Wilson, but in his case consilience is the — arguably Procrustean — fusion of all knowledge into a framework of scientific naturalism.

Creationists can also be subsumed under this model, although their integration is diametrically opposed to that of Wilson.

4. The complementarity model

This view owes something to the second model, arguing that science and religion do address different kinds of question but in a complementary fashion, producing different types or levels of mapping — as in systems theory. Alexander illustrates this approach with reference to the human body as understood by physics and biology. A putative super-scientist from Mars might additionally wish to know how humans arrived at ethical decisions and why there are conscious beings on our planet in the first place. Maps are not rivals but view the same reality through different lenses — various levels of description are necessary for an adequate overall picture.

Alexander then revisits the four types of model, asking how they related to human values. In this context the 'Conflict Model' looks unpromising as it tends to undermine rather than support concepts of human worth and dignity. The NOMA model fares rather better by insisting that science has little to contribute to discussions on moral and ethical issues. The geneticist Steve Jones is a typical representative when he maintains that science cannot ask 'why' or 'ought' questions. However, the alleged separation between facts and values cannot be as absolute as Jones assumes. Some areas of research are more ethically loaded than others and we need accurate scientific information for an informed ethical discussion. One variant of the integrationist model maintains that evolutionary understanding does generate an ethic, as in the case of Michael Ruse. Alexander devotes considerable space to analysis of Ruse's five-step process to argue from biology to ethics:

1. Complex human behaviours can be inherited.

2. Such behaviours have (or once had) adaptive value.

3. The force of 'ought' is in fact based on biological drives — 'morality a collective illusion foisted [on us] by our genes' according to Ruse.

4. Biological drives result in ethical impulses broadly in line with traditional morality

5. We have a moral duty to aid the process of evolution since it
   has generated such moral beliefs.

Alexander suggests that two types of critique can be levelled at
Ruse — empirical and philosophical. There is little experimental
support for steps 1 & 2 , as argued in the recent Nuffield Council
of Bioethics 'Genetics and Human Behaviour' report. And if
moral convictions do have adaptive value, this might equally
work through cultural transmission. Step 3 is even more problematic and runs aground in the exposition by the Cambridge
philosopher G.E. Moore. To the extent that 'ought' is redefined
as an innate biological disposition, the force of the concept
'vaporises', as Dr. Alexander colourfully expressed it: 'moral
obligations founded on a disposition to do something are not
really obligations at all'. Step 4 — the claim that biological drives
correspond to the lines of traditional morality — is empirically
false. People are not robots and do make genuine choices — there
is no real criterion of value. Indeed the *reductio ad absurdum* of
Ruse's position would be to describe atrocities as simply an
interesting manifestation of innate dispositions. Alexander concludes that we therefore have to be cautious about extrapolations
from genetics into values.

For Alexander the Complementarity Model fares rather better.
Here science is acknowledged to be restricted in the kind of
knowledge that it generates through quantification, measurement and description. Its body of knowledge gives us a precise
description of the physical world, but its descriptions necessarily leave out huge areas relating to ethics, aesthetics and questions of ultimate meaning and value. In the Complementary
Model meaning and purpose are important so as to do justice to
the fullness of experience. The model highlights the point that
the sources of human values do not arise from science. Alexander illustrates his point with reference to a controversial debate
on infanticide, arguing that a given metaphysical presupposition
can lead logically to an abhorrent conclusion, as is the case with
Peter Singer, who rejects a theistic framework in making his case
for infanticide, pointing out that Greeks and Romans commonly
killed deformed infants.

Our notion of the sanctity of human life derives from Christianity and infanticide was made illegal after the conversion of
the Emperor Constantine to Christianity in AD 313. If one asks

what Singer puts in the place of the Christian view, the answer is that he is a consequentialist — ethical decisions should be based on an assessment of their overall consequences on the basis of the 'principle of equal consideration of interests'. Singer extends this principle to sentient animals who can suffer, arguing that primates also self-conscious and can therefore be defined as 'persons' with interests.

The conclusion of such a starting point is that 'it is not intrinsically wrong to kill a newborn baby because the baby is not yet self-conscious whereas it would be wrong to kill adult animals who are supposed to be self-conscious'. Singer ends up by arguing that 'killing a disabled infant is not morally equivalent to killing a person. Very often it is not wrong at all'. The Christian position, by contrast, maintains that human life a gift from God and love and that the helpless new-born child has intrinsic value independent of biological status because humans are made in the image of God. We have a responsibility before God and are all part of the human community. There is a relationship and solidarity since all humans are God's image-bearers.

Alexander reminds us that moral decisions informed by Christian ethics are not necessarily black and white. Consequentialist arguments can be useful in deciding between the lesser of two evils, for instance in medical ethics. However, the metaphysical framework within which ethical reflection is carried out does make a real difference. So Alexander concluded that science can only retain a human face within a framework that sustains intrinsic human value.

John Habgood's lecture was given in Oxford under the related theme of Science and Human Responsibility. His paper is informed not only by his wide reading in theology and philosophy, but also by his involvement in ethical issues put before the House of Lords or government committees in which he has played a key role. He begins with the case of Diane Blood, a woman who conceived a second child using sperm from her deceased husband. Her action, he thinks, overstepped the normal boundaries between life and death and created a tension between sympathetic understanding of a particular individual case and its extrapolation into widespread social use. It is therefore important to think such matters through and to consider the long-term social consequences of a generalisation of such a technique. This is the function of the legislators, who have to ask themselves if the possible should become the permissible.

Habgood asks what the wide employment of such techniques implies for our understanding of ourselves as persons? It would give us the capacity to reproduce far into the future, creating 'orphans by design' and would arguably encourage competition for valuable genes, 'a car boot sale of genetic products'. Egg and sperm then become commodities and indeed have become so already on some web sites.

Habgood next considers identity (and dignity) in relation to adoption and to AI using third-party donors. In both cases there is a distinction between genetic and social parents, but in the case of AI the child might have 'an extensive network of half-siblings'. How might this emotionally affect the child's sense of self, and what role might the donor as an invisible third party play in the family dynamics? The point is that new techniques may have social repercussions on our idea of personhood and should be thoroughly thought through when framing legislation.

Habgood's next example is Tony Bland, a young man who had been in a permanent vegetative state (PVS) since his accident at the Hilsborough football stadium. Here the case involves the presence of a body without consciousness, while the previous one entails reproduction without a body. However, in neither case is consciousness embodied and both cases defy our normal understanding of limits. In Tony Bland's case his personhood was maintained by the daily visits of his parents, which defined a certain bodily relationship. But the parents eventually decided that there was no future in such a relationship and sought permission for his life to be terminated; something then died as a result of this decision, before it was actually ratified and carried through. Personhood, Habgood therefore contends, is crucially defined by our interaction and relationship with others. Although this relationship can transcend death, it is redefined by it, and demands a letting go on the part of the survivor. In the case of Diane Blood, she was arguably not letting go of her husband and was even using him (although with his prior consent before his death).

If this line of thinking leads to a prioritisation of genes over the body as a whole, it can make us forget that life events are more important than our genetic inheritance and that we must take the whole of what we are into account; and in the widest sense our environment includes God. It is the human personality, not the genes, that is the point of intersection between science and religion. And the focal point of human personality is the mystery of consciousness, our inwardness and sense of subjectivity. Habgood argues that consciousness is an 'impenetrable mystery' whose essential character cannot be observed or studied objectively, although we can on the other hand study thinking, feeling and brain mechanisms.

He then asks how subjective awareness has evolved, and attributes a key role to language in its capacity to create rapport by marking shared experiences and feelings. These in turn create social bonds and form community; and in this respect gossip has an important role to play. Through interchange with others we form ourselves, we discover both ourselves and others as agents. This takes us out of ourselves and helps develop moral imagination, locating our subjectivity in the larger context of a public world in a similar sense to that argued by Mary Midgley above.

Like Midgley, Habgood also discusses the role of environment, by which he meant all that impinges on us, for instance the social environment. The environment is not simply a backcloth but rather (as the Gaia hypothesis proposes) the scene of active engagement. Humans are 'uniquely interactive with the environment', which we adapt to our needs, including aesthetic satisfaction. The next question is where the environment ends and what our relationship might be to ultimate reality, however we understand the term. This must surely include subjectivity, but Habgood deliberately resists the temptation to short-circuit the process by bringing in God at this point.

Instead, he pursues his theme of language, pointing out that it extends beyond survival in its functions represented by art and theatre. We have an inherent yearning for transcendence: openness is built in, a sense of freedom, creativity and space. Habgood rejects panpsychism as unintelligible, preferring the notion of createdness. This meant that the existence of things was not self-explanatory. They (and we) exist by virtue of some reality that transcends us. The thesis of creation by God allows for rational exploration and implies a certain distance between the world and God whereby the world is allowed to make itself. This corresponds to the Incarnation, an acceptance of limitations and a renunciation of all power except love. The world is this sustained in being by God without God actually causing everything. Creation is not the outworking of a pre-ordained plan, hence science is important. God is distant but not absent, He transcends creation but can be known through it.

This view sheds light on subjectivity in a way that does not constitute proof of createdness, but which nevertheless makes sense, as Alan Torrance also argues. God can be known by those who seek. The development of personhood takes place in a context of interaction in freedom and dependence, of self-assertion and self-expression, for instance in loving and being loved. The possibility of relationship already exists through creation, which opens up the desire for deeper relationship. Recognising our dependence on God means recognising what we are at the deepest level of existence, where we can 'know as we are known'. Habgood feels that we bear the imprint of our transcendent origin.

In this context he brings in the 'I am' sayings in the Old and New Testaments. God is the 'Universal Subject of all experience', of all being and knowing. The sayings are the basic expression of subjective consciousness that is at once transcendent and incarnate. Our self-awareness is called into existence, we are known from within so that we can know ourselves as subjects. Creaturely dependence is a fundamental feature of existence, but a feature from which freedom, autonomy and self-awareness are drawn out as we are 'made in the image of God'. In a sense, Habgood maintains, science can share this quality of creativity but this does not mean that it can cross all boundaries or that we should envisage a future made in our own image — this is arguably the opposite of true development.

Concluding Observations

After a particularly intricate legal case, F.E. Smith (later Lord Chancellor as the Earl of Birkenhead) was asked by the judge to summarise the proceedings. At the end of Smith's disquisition, the judge commented that he was none the wiser — then at least, retorted Smith, you are better informed! Readers of this introduction may have some sympathy with the judge, although I hope also to have whetted the appetite for the papers that follow.

It seems to me that a number of overall themes have been thrown up by contemporary discussions about consciousness, especially given the legacy of the Western philosophical tradition, now less isolated from Indian and Chinese views than it was even fifty years ago. A prominent issue is how to deal with our inheritance of dualism: in philosophy and psychology this becomes a dualism between mind and matter, inner and outer; in ecology it is humans and the earth and, in social or ethical terms, self and other. One move — now widely discredited — is to try collapsing one into the other. Since the seventeenth century this has meant mind and consciousness progressively collapsing into matter, a position still held by many mainstream scientists and systematically laid out in E.O. Wilson's *Consilience.*

Our authors, however, are not content with this Procrustean solution. They reject both behaviourism and epiphenomenalism as incoherent, and attempt in their own ways to reassert the fundamental nature of consciousness. Few go as far as Fenwick and Goswami in asserting the total primacy of consciousness, but all are seeking to give it a more creative role in the scheme of things and to formulate a science of consciousness on a new basis. However, this does not gloss over different understandings of the self or self-system as Guy Claxton calls it. Is there an element of the human being that survives the death of the body? For materialists this is impossible by definition since mind = brain. If consciousness does indeed survive in some form then are we back to the dualism that we are trying to transcend? This may explain some of the resistance to considering evidence for what biologist Rupert Sheldrake calls 'the extended mind'.

In one sense the answer would have to be yes, since survival implies a consciousness separable from the body, as reported in NDEs. However there is a further dichotomy of western philosophy that needs to be overcome, namely that between embodied and disembodied existence where the meaning of embodied is restricted to the physical body. We saw in Keith Ward's piece that St Paul makes a distinction between the physical and spiritual bodies, while in Indian philosophy there is a tradition of subtle bodies. In his book *Paradigm Wars,* the American philosopher Mark Woodhouse suggests that we can envisage the co-existence of consciousness and energy/form (mind and body) at a number of different levels. We know what this means at the physical level, but Woodhouse's view implies that a putative post-mortem existence might (initially at least) involve the same consciousness in a different, more 'subtle' body, a theory already supported by the widespread experience of apparitions.

So I agree with Peter Fenwick when he asserts that a new science of consciousness must address these (currently anomalous) phenomena.

Discussion of subtle bodies is not, however, central to the transformation of consciousness enjoined by Ravi Ravindra and Guy Claxton. This process involves contemplative practices designed to train the attention to see and perceive more directly, with less personal and cultural conditioning to cloud the picture. It is personal work that can only be undertaken at an individual level and which is necessary if new scientists of consciousness are to come to a deeper understanding of their own minds, reaching beyond the ego to what Ravindra calls the First Person Universal. The work of the Buddhist philosopher Alan Wallace (*The Taboo of Subjectivity*) is seminal in this regard.

Wallace's approach addresses the imbalance in Western culture noted in David Fontana's paper and perhaps points to the cultural significance of the emergence of a science of consciousness in our time. If investigations are confined to the third-person perspective, then little will change. We will remain fixed in that outer orientation that largely pervades science, psychology and philosophy and which largely neglects the significance of subjective experience. However, to the extent that we engage in systematic investigation of inner states, we will find our understanding gradually transformed in the process. Not only will we have a new, more participatory science of consciousness, we will also have self-aware scientists open to deeper aspects of reality.

Bernard Carr

Mind and the Cosmos

Introduction

The purpose of this paper is to consider whether mind is a fundamental or incidental feature of the cosmos. The first part will describe how the progress of science has entailed a continual change in our view of the Universe and a constant retreat from the anthropocentric perspective of early Man. We seem to have become increasingly insignificant and in many ways mind now appears irrelevant to the functioning of the Universe. However, in recent decades there has been a reversal of this trend and the second part of the paper will describe the various indications from science itself that mind may be significant after all. This also bears on the science-religion debate; science cannot prove or disprove the existence of a divine element in the cosmos, but the miracle of matter may provide just as strong an intimation of the divine as the miracle of mind. The third part of the paper, while very speculative, suggests that the next scientific paradigm may incorporate consciousness in a very fundamental way by invoking an extended view of reality which unifies matter and mind explicitly. Recent developments in theoretical physics and psychical research may already provide a glimpse of what form this paradigm may take.

PART 1: SCIENTIFIC PROGRESS AND THE DEMEANING OF MAN

Throughout this paper the term 'cosmos' will be used more or less interchangeably with the term 'Universe' to mean the entirety of physical creation. The term 'mind' will be used very loosely, being applied variously to describe Man, consciousness and spirit. We will start by tracing how scientific progress has continuously changed our view of the cosmos and of the role of mind within it.

Our Changing View of the Cosmos

Early Man started with the 'geocentric' view in which the Earth was located at the centre of the Universe. The world was very much 'alive' and Man was the focus of creation, with a direct link to the guiding intelligence which created and sustained the world. Astronomical events were interpreted as being much closer than they actually are, because the heavens were assumed to be the domain of the divine and therefore perfect and unchanging. Even the laws of Nature (such as the regularity of the seasons) seemed to be Man-centred, in the sense that they could be exploited for our own purposes, so it was natural to regard them as a direct testimony to our central role in the Universe.

However, this 'anthropocentric' perspective was shattered once science started to expand its domain of interest beyond the human scale. By developing new instruments, like the telescope and the microscope, it was possible to extend observations *outwards* to scales much bigger than Man and *inwards* to much smaller scales. New levels of structure and new laws were thereby found and, wherever Man looked, there was evidence for the same sort of regularity as had been found at the mundane level. There were also interesting connections between the laws perceived at the different levels. However, while the journey of discovery has been intellectually gratifying, it has also been psychologically painful. For as Man has gained knowledge, so his perspective of his place within the cosmos has shifted and the shifts have nearly always reduced his status. Often the new perspective has clashed with the conventional (non-scientific) dogma of the day, leading to conflict with religious orthodoxy.

The demeaning process was initiated during the Renaissance period by three crucial steps on the outer front. In 1543 Nicolaus Copernicus showed that the heliocentric picture provides a simpler explanation of planetary motions than the geocentric one, thereby removing the Earth from the centre of the Universe. Then in 1572 Tycho Brahe spotted a supernova in the constellation of Cassiopeia: it brightened suddenly and then dimmed over the course of a year but the fact that its apparent position did not change as the Earth moved around the Sun implied that it was well beyond the Moon. This destroyed the Aristotelian view that the heavens never change. Then in 1610 Galileo used the newly invented telescope to remove the specialness of the Sun: his observations of sunspots showed that the Sun changes and he also speculated that the Milky Way consists of stars like our Sun.

A more subtle development was Newton's discovery of universal gravity. By proposing that the force which makes the apple fall to the ground also holds the planets in orbit around the Sun, Newton connected heaven and Earth and thereby removed the special status of heaven. Furthermore the publication of his *Principia* in 1687 led to the 'mechanistic' view, in which the Universe is regarded as a giant machine. For Newton himself this testified to the need for God and thereby, implicitly, to the importance of Man:

> Blind fate could never produce the wonderful uniformity of planetary movements. Gravity may put the planets into motion but without the divine power, it could never put them into such circulating motions as they have.

However, this blend of science and theism was not to persist with most of his successors. Mechanism was soon stripped of its divine aspects and before long Man was generally perceived as being completely irrelevant to the functioning of the Universe.

The Outward Journey

Since the Renaissance, advances on the outer front have reduced the status of Man still further. The development of ever more powerful telescopes has enabled us to peer to ever greater distances. Not surprisingly, the further we have looked, the bigger the Universe has become and the more Man has shrunk by comparison. Galileo's realisation that the Milky Way is composed of other suns had already vastly increased the size of the Universe, since they had to be at enormous distances. However, it was several more centuries before astronomers began to understand the structure of the Milky Way and to realise that the Sun is just one of a hundred billion stars which comprise the Galaxy. Although early models placed the Sun at the centre of the Galaxy, this turned out not to be the case. Nevertheless, since it was assumed that the Milky Way comprised the whole Universe, one could still adopt a Galactocentric view. But then in the 1920s it was realised that some of the nebulae observed by astronomers are 'island universes' outside the Milky Way. These objects turned out to be galaxies just like our own, and we now know that the observable Universe contains a hundred billion of them.

An even more dramatic revelation came in the late 1920s when Edwin Hubble discovered that all galaxies are moving away from us. The most natural interpretation of this is that space itself is expanding — this was indeed predicted by Einstein's theory of General Relativity — with all galaxies receding under the impetus of a Big Bang which occurred ten billion years ago. So, as if the Universe was not big enough already, it now transpired that it was getting bigger all the time. Not only the Sun but the cosmos itself is continuously changing!

In 1965 astronomers discovered some striking evidence for the Big Bang picture. They found that the Universe is bathed in a sea of background radiation, which is thought to be a relic of when the Universe was only a million years old. Subsequent studies of this radiation, most recently by satellites such as COBE and WMAP, show that it has a perfectly black-body spectrum — confirming its Big Bang origin — and also reveal the tiny temperature fluctuations associated with the ripples which eventually led to the formation of galaxies and clusters of galaxies. George Smoot, the principal investigator of the COBE project, described the picture of these ripples as the 'face of God' and one of the prime aims of cosmologists today is to study this 'face' in ever greater detail.

The Inward Journey

In Man's probing of the smaller scales, our demeaning and the clash with orthodoxy was at first less pronounced, primarily because religious dogma had little to say on the matter. Nevertheless, with the advent of atomic theory came the first hints that

Man's experience of the small is just as limited as his experience of the large. The discovery that all matter is composed of atoms, with consequent insights into chemistry, was not itself disturbing. So long as they could be viewed as solid objects, like billiard balls, they more or less complied with our intuition of how objects in the real world ought to behave. However, there was also a cloud on the horizon, because atomic theory was linked to statistical mechanics and the second law of thermodynamics suggested that the Universe must eventually undergo a 'heat death', with life (including Man) and all other forms of order inevitably deteriorating.

Further dramatic developments came early in the twentieth century. The 'billiard ball' picture of the atom was demolished by Ernest Rutherford, who showed that an atom is mainly empty space, with electrons in orbit around a nucleus comprised of protons and neutrons. It must have been unsettling to find the real world deprived of solidity like this, but at least the constituents of the atoms could be regarded as ordinary particles. However, soon even this solace was removed with the discovery of quantum theory. This suggests that elementary particles no longer behave like billiard balls. Rather they are fuzzy, ephemeral entities, described by a 'wave function' which is smeared out everywhere. The classically deterministic laws of cause and effect were replaced by probabilistic laws, in which events previously considered impossible would occur routinely.

So the discovery of quantum mechanics shattered Man's perspective of the microworld just as much as the discoveries of astronomy shattered his perspective of the macroworld. Man's probing of the microworld also emphasised his vulnerability. For he discovered new forces — the strong force (which holds the nucleus together) and the weak force (which controls radioactive decay) — and thereby unleashed an awesome new source of destruction. Atoms might be smaller than us but they evidently contained powers which made our own seem puny. We were forced to realise that we are just as prone to destruction from the world within as from astronomical perils.

Despite this, the inner journey has reaped huge intellectual rewards: it has revealed that everything is made up of a few fundamental particles and that these interact through four forces: gravity, electromagnetism, the weak force and the strong force.

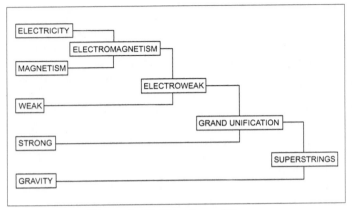

Figure 1. Unification of Forces

These interactions have different strengths and characteristics and it used to be thought that they operated independently. However, it is now thought that some (and possibly all) of them can be unified as part of a single interaction. Indeed the history of physics can be seen as the history of this unification, as illustrated in Figure 1. Electricity and magnetism were combined by Maxwell's theory of electromagnetism in the nineteenth century. The electromagnetic force was then combined with the weak force in the (now experimentally confirmed) 'electroweak' theory of the 1970s. Theorists have subsequently merged the electroweak force with the strong force as part of a Grand Unified Theory (GUT), although this has still not been verified experimentally. The final (and as yet incomplete) step is the unification with gravity, as attempted by superstring theory or M-theory. Indeed many people have proclaimed that the end of physics is in sight, in the sense that our knowledge of the fundamental laws and principles governing the Universe is nearly complete. They argue that we are on the verge of obtaining a 'Theory of Everything'.

The Uroborus and the Big Bang

Taken together, scientific progress on both the outer and inner fronts can certainly be regarded as a triumph. In particular, physics has revealed a unity about the Universe which makes it clear that everything is connected in a way which would have seemed inconceivable a few decades ago. This unity is succinctly

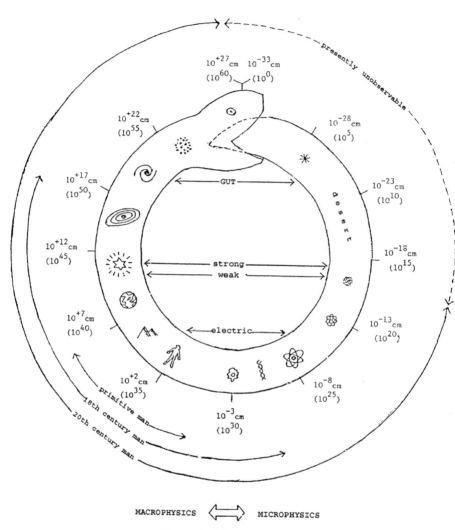

Figure 2. The Uroborus

encapsulated in the image of the snake swallowing its tail (the *Uroborus*). This is shown in Figure 2 and demonstrates the intimate link between the macroscopic domain (on the left) and the microscopic domain (on the right). The pictures drawn around the snake represent the different types of structure which exist in the Universe. Near the bottom is Man himself. As we move to the left of Man, we encounter successively larger objects: a mountain, a planet, a star, a solar system, a galaxy, a cluster of galaxies and finally the entire observable Universe. As we move to the right, we encounter successively smaller objects: a cell, a DNA molecule, an atom, a nucleus, a quark, the GUT scale and finally the Planck length (the scale at which quantum gravity effects become important).

The numbers at the edge indicate the scale of these structures in centimetres. As one moves clockwise from the tail to the head, the scale continuously increases, so the body of the snake provides a sort of ruler. The tip of the tail corresponds to the Planck length (10^{-33}cm) and the head corresponds to the size of the observable Universe (about 10^{27}cm). In Planck units, the scales (in parentheses) go from 10^0 to 10^{60}, so the Uroborus is like a clock, with each minute corresponding to a factor of 10 in scale.

A further aspect of the Uroborus is indicated by the horizontal lines. These correspond to the four interactions (discussed above) and illustrate the subtle connection between the microphysical and macrophysical domains. For example, the 'electric' line connects an atom to a mountain because the structure of a solid object is determined by atomic and intermolecular forces, both of which are electrical in origin. The 'strong' and 'weak' lines connect a nucleus to a star because the strong force which holds nuclei together also provides the energy released in the nuclear reactions which power a star and the weak force which causes nuclei to decay also prevents stars from burning out too soon. The 'GUT' line connects the grand unification scale with galaxies and clusters because the density fluctuations which made these objects originated when the temperature of the Universe was high enough for GUT interactions to be important. Indeed the Big Bang theory suggests that all features of Universe arose when the current observable Universe had the size of a grapefruit!

The significance of head meeting tail is that the entire Universe was once compressed to a point of infinite density. Since light travels at a finite speed, the further out we look in space, the further back we are looking in time. In fact, we can never see further than the distance light has travelled since Big Bang, about 10^{10} light-years, so more powerful telescopes merely probe to earlier times. Cosmologists now have a fairly complete picture of the history of Universe: galaxy formation occurred at a billion years after the Big Bang, the background radiation last interacted with matter at a million years, the Universe's energy was dominated by its radiation content before about ten thousand years, light elements were generated through cosmological nucleosynthesis at around three minutes, antimatter was created at about a microsecond (before which there was just a tiny excess of matter over antimatter), electroweak unification occurred at a billionth of a second (the earliest time which can be probed experimentally using accelerators), grand unification and 'inflation' (an extra-rapid expansion phase) occurred at 10^{-35}s, and the quantum gravity era was at 10^{-43}s.

Arguments for the Insignificance of Mind

At this point in the story, the various branches of science all appear to collude to diminish the status of Man. The heavens have been stripped of consciousness and the more we understand the Universe, from the vast expanses of the cosmos to the tiny world of particle physics, the more soulless and inanimate it seems to become. The religious view that we have a special significance has been shattered. As Steven Weinberg says: 'The more the Universe seems comprehensible, the more it seems pointless.' Let us summarise the reasons for gloom.

- *Man is insignificant*

The steady progress from the geocentric to heliocentric to galactocentric to modern cosmic view shows that Man — as judged by scale — is incredibly insignificant. Indeed recent developments suggest that our entire Universe may just be one member of an ensemble of universes — called the 'multiverse'. Man is equally insignificant as judged by duration: the lifetime of an individual man — and even an entire civilisation — is utterly negligible compared to the time scale on which the cosmos functions. If the history of the Universe were compressed to a year, *homo sapiens* would have persisted for only a few minutes. Nor is it clear how long Man will persist in the future, since we are prone to dangers from both without (asteroids, marauding black holes, exploding stars) and within (some devastating new virus). Today the futility of existence is accentuated by the possibility that humans will eliminate themselves and most other life

forms on the planet in a nuclear holocaust or some other environmental catastrophe. Everything in the Universe falls apart and
the presence of Man, it seems, merely accelerates the process!

• *Life is result of chance processes*

Although we have focussed on the physical sciences, biological
advances have been equally humbling. Darwin's theory of evolution showed that *homo sapiens*, far from being a unique creation
made in God's image in the Garden of Eden, is just the latest
stage of development in a long series of biological mutations. For
a while, one could still think of God as guiding evolution or at
least initiating it. However, with the discovery in the 1950s of
DNA, the complex molecule containing the genetic information
on which heredity is based, it became clear what steers evolution
is not some divine being outside but a tiny molecular strand
inside.

• *The Universe — including mind — is just a machine*

Since the Enlightenment, the prevailing scientific view has been
that the Universe — and everything in it — is just a machine.
Indeed, every technological innovation is based on this assumption. Recent advances in brain research and artificial intelligence
suggest that even the mind is a machine. We may appear to have
free will but this could just be an illusion, consciousness being
the mere excretion of a brain that is itself a machine. We can
already envisage machines that think more quickly, remember
more precisely and decide more intelligently than mere human
minds. The mechanistic outlook also seems to have done away
with the need for any divine element in the world, so much so
that Richard Dawkins can now dismiss believers in a creator as
'scientifically illiterate'.

• *The Big Bang removes the need for a divine creator*

The Big Bang theory does not itself preclude God since one can
still ask 'who lit the fuse'. Indeed the fact that the Universe had a
finite beginning was regarded by Pope Pius XII as supporting
the Genesis picture. Implicit here is the notion that the physical
description of creation is incomplete, since it must break down at
sufficiently early times. However, as time proceeds, physics
seems to have provided an ever more complete description. For
example, one could envisage the following dialogue:

How did the Universe originate? The Universe started as a state of compressed matter. But where did the matter come from? The matter arose from radiation as a result of GUT processes occurring when the Universe had the size of a grapefruit. But where did the radiation come from? The radiation was generated from empty space as a result of a vacuum phase transition. But where did space come from? Space appeared from nowhere as a result of quantum gravity effects. But where did the laws of quantum gravity come from? The laws of quantum gravity are probably no more than logical necessities.

Each step in this dialogue represents many years of painstaking theoretical work but the upshot of the argument is clear. No first cause is needed because the Universe contains its own explanation. That, for example, is the view propounded in Stephen Hawking's book 'The Universe in a Nutshell'. Even if God does exist, it is not clear He could have created the Universe differently. So unnecessary is He that scientists now speculate about creating universes in the laboratory!

PART 2: THE REVIVAL OF MIND

Although the advance of science seems to have systematically diminished the significance of mind, recent decades have seen a curious reversal in this trend. The reversal has come from science itself and stems from a number of factors: an indication of intelligence and beauty in the laws of Nature; the increasing credence given to anthropic arguments; the realisation that the presence of mind is the natural culmination of an organisational principle in the Universe; and an appreciation that the picture of reality provided by science is itself just a mental model. We now describe these developments in more detail.

The Other Side of the Uroborus

Although the triumph of physics — which appears to demean Man so much — is encapsulated in the Uroborus, various other features of the Uroborus actually point to the opposite conclusion. For example, it is striking that Man occupies a special symmetry point near the bottom, midway between the largest macrophysical scales and the smallest microphysical ones. This reflects the fact that our size is roughly the geometric mean of the

Planck length and the size of the observable Universe. As Protagoras said: 'Man is the measure of all things.'

The Uroborus can also be interpreted historically as representing the evolution of consciousness. For the outer arcs show how Man, through scientific exploration, has systematically expanded the outermost and innermost limits of his perception. Primitive Man was only aware of the mundane scales between about 10^{-1}cm and 10^{7}cm. The invention of the first microscopes and telescopes then expanded the range from about 10^{-5}cm to 10^{17}cm. Today the development of particle accelerators and space telescopes has expanded the range from 10^{-15}cm to 10^{27}cm. Indeed, in terms of scale, it is striking that science has already expanded the macroscopic frontier as far as possible, although we may never get much below the electroweak scale in the microscopic direction.

The unity of creation and the intimate link between the macroscopic and microscopic expressed by the Uroborus has also led some scientists to see evidence of a great intelligence at work in the Universe. For example, James Jeans suggested that 'The Universe is more like a great thought than a great machine.' This impression derives from the fact that the Universe is so cleverly constructed. At the very least, the coherence of the laws which regulate it seems to point to the existence of some underlying organising principle. This also relates to the question of why the Universe is comprehensible at all. It seems remarkable that after just a few millennia, we are already on the verge of a Theory of Everything. As Roger Penrose emphasises, there seems to be a closed circle: the laws of physics lead to complexity, complexity culminates in mind, mind leads to mathematics, and mathematics allows an understanding of physics.

There is also an inherent beauty in the Universe. The nature of this beauty is hard to define but it involves mathematical elegance, simplicity and inevitability (like good music, it cannot be modified even slightly without destroying whole). In particular, all the laws of Nature seem to be a consequence of a simple set of symmetry principles. For example, symmetrising electricity and magnetism gave Maxwell's equations; symmetrising space and time gave Special Relativity; and invoking gauge symmetries has led to the unification of the forces of Nature. Such symmetries can only be appreciated intellectually but they are profoundly elegant and can be very moving for physicists. The importance of beauty was appreciated by Paul Dirac, who claimed 'Beauty in equations is more important than fitting experiments,' and by John Wheeler, who said 'One day a door will surely open and expose the glittering central mechanism of the world in all its beauty and simplicity.' Since intelligence, comprehensibility and beauty are attributes of the mind, this suggests that mind is not a purely incidental feature of the world.

The Anthropic Principle

We have seen that the progress of science rapidly led to the over-throw of the anthropocentric view and rise of the mechanistic view. In the last 40 years, however, there has developed a reaction to mechanism which is called the 'anthropic' view. This claims that, in some respects, the Universe has to be the way it is because otherwise it could not produce life and we would not be here speculating about it. Although the term 'anthropic' derives from the Greek word for 'Man', it should be stressed that this is really a misnomer since most of the arguments pertain to life in general rather than Man in particular. A better description might be the 'complexity' principle.

As a simple example of an anthropic argument, consider the question: Why is the Universe as big as it is? The mechanistic answer is that, at any particular time, the size of the observable Universe is the distance travelled by light since the Big Bang. Since the Universe's present age is about 10^{10}y, its present size is about 10^{10} light-years. Inherent in this straightforward answer is the belief that there is no compelling reason the Universe has the size it does; it just happens to be 10^{10}y old. There is, however, another answer to this question, one which Robert Dicke first gave in 1961. His argument runs as follows: in order for life to exist, there must be carbon or at least some form of chemistry. Now carbon is produced by cooking inside stars and this process takes about 10^{10}y. Only after this time can the star explode as a supernova, scattering the newly-baked elements throughout space, where they may eventually become part of life-evolving planets. On the other hand, the Universe cannot be much older than 10^{10}y, else all the material would have been processed into dead stellar remnants. Since all the forms of life we can envisage require the existence of stars, this suggests that life can only exist when the Universe is aged about 10^{10}y.

This startling conclusion turns the mechanistic answer on its head. The very hugeness of the Universe, which seems at first to point to Man's insignificance is actually a prerequisite of his existence. This is not to say that the Universe itself could not exist with a different size, only that we could not be aware of it when its size was different. Of course, it could just be a coincidence that the age of the Universe happens to be about the time required to produce intelligent life. However, this argument does at least give a taste for the type of reasoning the Anthropic Principle entails. Indeed the evidence for the Anthropic Principle rests almost entirely on the large number of numerical 'coincidences' in physics which seem to be prerequisites for the emergence of life and which would otherwise have to be regarded as purely fortuitous.

Dicke's argument is an example of what is called the 'Weak Anthropic Principle'. This accepts the constants of nature as given and then shows that our existence imposes a selection effect on the epoch at which we observer the Universe. As such, it is no more than a logical necessity: saying that we have to exist at a certain time is no more surprising than saying that we have to live in a certain place (e.g. close to a star). Much more controversial is the 'Strong Anthropic Principle', which says that there are connections between the coupling constants (the dimensionless numbers which characterise the strengths of the four interactions) in order that life can arise. For example, the existence of stars with planets requires that there be a tuning between the electric and gravitational coupling constants; heavy elements like carbon can be ejected from stars in supernova explosions only because there is a tuning between the weak and gravitational coupling constants; and an interesting variety of chemical elements exists only because there is a tuning between the electric and strong coupling constants. These relations correspond to vertical connections in the Uroborus diagram.

Various sorts of interpretation have been suggested for the anthropic coincidences and these are illustrated in Figure 3. The first possibility is that they reflect the existence of a 'beneficent being' who tailor-made the Universe for our convenience (Figure 3a). Such an interpretation is logically possible, and

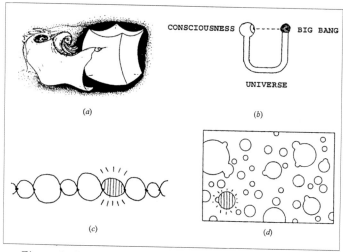

Figure 3. Interpretations of the Anthropic Principle

obviously appeals to theologians, but it is unpalatable to most physicists.

Another possibility, proposed by John Wheeler, is that the Universe does not properly *exist* until consciousness has arisen. This is based on the notion that the Universe is described by a quantum mechanical wave function and that consciousness is required to collapse this wave function. Once the Universe has evolved consciousness, one might think of it as reflecting back on its Big Bang origin, thereby forming a closed circuit which brings the world into existence (Figure 3b). Even if consciousness really does collapse the wave function (which is far from certain), this explanation is also somewhat metaphysical.

The third possibility is that there is not just one Universe but lots of them, all with different randomly-distributed coupling constants. In this 'multiverse' proposal, we just happen to be in one of the small fraction which satisfy the anthropic constraints. With this interpretation, the fact that the constants have the values required for life just becomes an aspect of the Weak Anthropic Principle. For conscious observers need not only be at special points in time and space but also in special universes. Invoking lots of extra universes might seem rather uneconomical but there are various physical contexts in which this possibility arises naturally.

One such context applies if the Universe contains enough matter to eventually recollapse (which is possible but uncertain). In this case, one could envisage it undergoing cycles of expansion and recollapse, with the coupling constants being reprocessed at every bounce (Figure 3c). During most cycles the constants would not allow life to arise but occasionally the appropriate values would occur and so the Universe would become aware of itself. However unlikely the constants are to have the required values, with an infinite number of cycles it is bound to happen sometimes.

Another 'multiverse' scenario arises in the context of the inflationary picture. This proposes that at very early times quantum fluctuations cause tiny regions to undergo an exponential expansion phase. Each region becomes a 'bubble' and our entire visible Universe is then contained within one of these. In principle, as stressed by Andrei Linde, there could be different values for the coupling constants within each bubble (Figure 3d). In this case, only a small fraction of them would develop consciousness but our Universe would necessarily be among that fraction.

Both the cyclic and inflationary models invoke a large — and possibly infinite — number of universes. The only difference is that the first invokes an infinity in time, while the second invokes and infinity in space. It must be stressed that all these explanations are very speculative but the point is that the 'multiverse' proposal does at least provide some basis for the Anthropic Principle. Admittedly, both pictures are probably untestable and, in this sense, one might still regard then as being metaphysical. Nor does the 'multiverse' explanation necessarily exclude God, since He could presumably create many universes as easily as one!

Perhaps the least radical explanation of the anthropic coincidences is that they may turn out to be a consequence of some unified theory of particle physics. Such theories do, after all, set out to relate the different coupling constants, so it is not inconceivable that they would predict the sort of connections between the coupling constants discussed above. However, as far as we know, the relationships discussed above are *not* predicted by any unified theory. Even if they were, it would still be remarkable that the theory should yield exactly the coincidences required for life.

It should be stressed that not all physicists accept the Anthropic Principle. Indeed they are very polarised about it, so we will end with some quotations. One is from the protagonist Freeman Dyson:

> I do not feel like an alien in this Universe. The more I examine the Universe and examine the details of its architecture, the more evidence I find that the Universe in some sense must have known we were coming.

This might be contrasted with the view of Heinz Pagels:

> The influence of the anthropic principle on the development of contemporary cosmological models has been sterile. It has explained nothing and it has even had a negative influence. I would opt for rejecting the anthropic principle as needless clutter in the conceptual repertoire of science.

An intermediate stance is taken by Brandon Carter, who coined the phrase 'anthropic principle', and this perhaps represents a sensible compromise:

> The anthropic principle is a middle ground between the primitive anthropocentrism of the pre-Copernican age and the equally unjustifiable antithesis that no place or time in the Universe can be privileged in any way.

The Pyramid of Complexity

We have seen that in the nineteenth century the second law of thermodynamics led to the pessimistic notion of 'heat death', with all forms of order − including life − inevitably decaying. However, according to the Big Bang theory, the history of the Universe reveals an increasing rather than deteriorating degree of organisation, and modern physics − without any recourse to divine intervention and without any violation of the second law of thermodynamics − is able to explain this. Heat death is avoided because local pockets of order can be purchased at the expense of a global increase in entropy (usually in the form of radiation) and, if the Universe continues to expand forever, intelligent beings may be able to delay their disintegration indefinitely.

Some of the types of organisation which exist in the Universe are summarised in Figure 4, which is adapted from Hubert Reeves' *Pyramid of Complexity*. This shows the different levels of structure as one goes from quarks to nucleons to atoms to simple molecules to biomolecules to cells and finally to living organ-

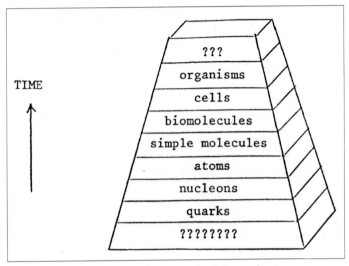

Figure 4. The Pyramid of Complexity

isms. This hierarchy of structure reflects the existence of the strong force at the lower levels and the electric force at the higher ones. As one ascends the pyramid, the structures become more complex — so that the number of different patterns becomes larger — but they also become more fragile. Note that the pyramid becomes narrower as one rises and this reflects the fact that smaller fractions of matter are incorporated into objects with greater degrees of organisation.

To understand *when* these structures arise, one must look to the Big Bang theory. This is because the Pyramid of Complexity only emerges as the Universe expands and cools. At early times the Universe is mainly in the form of quarks. Neutrons and protons appear at a few microseconds, light nuclei at several minutes, atoms at a million years, and — following the formation of stars and planets — molecules and cells at ten billion years. The Big Bang theory also explains *why* the pyramid came about. The key point is that structures arise because processes cannot occur fast enough in an expanding Universe to maintain equilibrium. All variety would be lost if equilibrium applied: the strong force would turn all nuclei into iron, the electric force would turn all atoms into inert gases and gravity would turn all matter into black holes. This disequilibrium is only possible because of the anthropic fine-tuning of the coupling constants discussed earlier. We are therefore midway between uniformity of the past and that of the future (when all matter will have decayed through quantum effects).

There is an important difference between the structures that exist at the top and the bottom of the pyramid. Those at the bottom are stable and need large amounts of energy to destroy them, while those at the top must be constantly maintained by exchanging energy with the outside world. More precisely, they must extract *information* from the world, and the second law of thermodynamics requires that this process is necessarily accompanied by the release of entropy (i.e. waste heat). A store of information arises whenever there exists a source of entropy that has not been released by previous processes. For example, nuclear information is contained in nuclei other than iron and this can be extracted by nuclear burning inside the Sun, the ultimate source and sustainer of all life forms on Earth. Similarly, living organisms can feed on plants and humans can exploit fossil fuels because these contain complex molecules with consumable electromagnetic information.

An important ingredient at the top level of the pyramid is *competitiveness*. This is a vital factor in evolution because, as a population grows, the competition for food leads to predation and increasingly sophisticated survival strategies. The proliferation of life forms due to mutation plays a crucial role in this process and sex - with its ability to mix gene pools - provides an essential source of diversity. Different modes of perception and motor activity are also required and this leads to the development of organisms with a central nervous system. From this perspective, brains — certainly the most complex structures on Earth — are merely data integration systems and the main purpose of intelligence is to increase survival efficiency.

Minds might be regarded as the ultimate storers and extractors of information, so their development can be seen as the inevitable culmination of the ascent of the pyramid of complexity. (Even if one does not adopt the reductionist view that minds are generated by brains, one presumably needs this ascent if minds are to operate within the physical world.) It is remarkable that — at least on Earth — the development of brains seems to have occurred relatively quickly once the first signs of life arose: the

Earth formed 4.5 billion years ago, bacteria then appeared after 0.5 billion years, self-replicating cells 2 billion years later, invertebrates after another 1.5 billion years, mammals some 300 million years after that, and finally *homo sapiens* after another 200 million years (just few million years ago). Therefore the Anthropic Principle — even if reinterpreted as a Complexity Principle — might still be regarded as attesting to the importance of mind.

Paradigm Shifts and Mental Models

The history of science shows that the prevailing model of physical reality regularly undergoes paradigm shifts. The paradigm determines the sort of picture one has of the world, the type of questions one asks about it and the experiments one performs. Much scientific progress is made within one particular paradigm but eventually anomalies arise and these result in a crisis which ultimately leads to the adoption of a new one.

The first paradigm was the classical *Newtonian* one, in which the physical world is regarded as a 3-dimensional continuum in which solid objects move according to Newton's laws of dynamics. Time is absolute, in the sense that it flows at the same rate for everyone, and there is also an absolute space associated with inertial (non-accelerating) frames. Objects attract each other through the force of gravity, although the paradigm does not explain *why* that force exists.

The next paradigm, *atomic theory*, arose from developments in statistical physics and thermodynamics. These showed how the interactions of billions of atoms naturally lead to the observed macroscopic laws and how the structure of the atoms themselves provides an understanding of chemistry. The new paradigm also contained the laws of electricity and magnetism. In particular, it showed that light consists of electromagnetic waves travelling through an 'ether', which was naturally identified with Newton's absolute space.

The advent of the third paradigm, *Special Relativity*, demolished the idea of the ether and showed that space and time are not absolute but part of a spacetime continuum. Thus a consistent picture of how different observers perceive the world requires that it be 4-dimensional rather than 3-dimensional, the fourth dimension being time.

The next transformation came with *General Relativity*, which showed that — in the presence of matter — spacetime is curved like a surface in a higher dimensional space. This explains the origin of gravity but gives slightly different predictions from Newton's theory. General Relativity also forms the basis of cosmology, the revellations of which were discussed in Part 1.

Paralleling these developments in macroscopic physics was the paradigm shift associated with *quantum theory*. This showed that microscopic objects can simultaneously behave like waves and particles. Measurements always interfere with objects in some way and this leads to the Uncertainty Principle. In particular, a particle cannot simultaneously be ascribed a position and velocity, which means that the worldline description of relativity can only be an approximation.

The *Kaluza-Klein* paradigm arose out of attempts to give a geometrical explanation of electromagnetic interactions, analogous to the geometrical explanation of gravitation provided by General Relativity. Kaluza-Klein theory suggested that the Universe is 5-dimensional; the fifth dimension is wrapped up so small that it cannot be observed directly but its existence neatly explains the laws of electromagnetism.

Modern extensions of this idea propose that the other interactions can be accounted for by invoking further wrapped-up dimensions. In the 'supergravity' theory, for example, the total number of dimensions is 11, while in the 'superstring' theory it is 10. One thus has a 4-dimensional external space and a 6 or 7-dimensional internal space. There are various versions of superstring theory but the most recent development suggests that these are all part of a more embracing picture called *M-theory* (where M stands for 'mother' or 'magic' or 'mystery').

The final — and as yet incomplete — paradigm shift is associated with *quantum gravity*, the attempt to unify General Relativity and quantum theory. According to this paradigm, the notion of space breaks down on scales less than 10^{-33} cm. It must be regarded not as a smooth continuum but as a sort of topological foam. Quantum gravity effects must dominate whenever classical physics predicts 'singularities' (i.e. points of infinite density).

This brief history of paradigm shifts shows that the 'ultimate reality' revealed by modern physics is very different from the sort of reality experienced by our usual senses, which only provide a very incomplete picture of reality. Indeed one can regard successive paradigms as providing a sequence of mental models, each of which is progressively removed from common-sense reality. Thus atomic theory removes our everyday notion of solidity, relativity theory destroys our intuitive ideas of space and time, quantum theory shows that reality is fuzzy, unification theories reveal dimensions of which we have no direct experience and quantum gravity goes beyond space and time altogether. It therefore seems that the ultimate nature of reality can only be appreciated intellectually, so it is ironic that many physicists wish to play down the significance of mind!

Arguments for the Significance of Mind

To counterbalance the points made at the end of Part I, we now summarise these arguments.

• *Man — or at least complexity — is central*

The Uroborus symbolises the triumph of physics but it also shows that Man plays a central role in creation. He is not at the centre of the Universe geographically but he is central in the scale of things. The significance of Man is also emphasised by the Anthropic Principle. Of course, these arguments do not apply exclusively to *homo sapiens*. They would apply to any other life-form which may have evolved in the Universe. Actually, the very bottom of the Uroborus corresponds to 10^{-3}cm, so perhaps the prestigious symmetry point should be associated with microbes rather than Man!

• *Life is more than an accident*

Life is an essential feature of the Pyramid of Complexity and several people have emphasised that there seems to be a 'life principle' at work in the cosmos. Astronomers still do not know how prevalent life is in the Universe. It seems unlikely that we are the only evolved form of life but attempts to discover extraterrestrial signals have not yet met with success. Whatever the outcome of these searches, the status of life is crucial. If Man turns out to be the only intelligent life in the Universe, then we are restored to our pre-Copernican status. If there are many of sites for life, then we are part of an evolving Galactic or even cosmic ecosystem.

• *Mind is fundamental to the cosmos*

The Uroborus also represents the blossoming of consciousness. The physical evolution of the Universe from the Big Bang (at the top) through the Pyramid of Complexity to Man (at the bottom) is just the start of a phase of *intellectual* evolution, in which mind - through scientific progress - works its way up both sides to the top again (cf. Figure 3b). In fact, this blossoming of consciousness corresponds precisely to the series of paradigm shifts described earlier. This also links to the question of extraterrestrial life. If intelligent life is widespread in the Universe, then it seems probable that mind has a hierarchical structure, with individual consciousness being part of planetary consciousness, which is itself part of Galactic consciousness etc.

• *The unity and beauty of the Universe point to a guiding intelligence*

We have seen that modern physics has revealed an unexpected unity and beauty in Nature, which seems to point to some underlying intelligence. It is natural (though not obligatory) to link this to the notion of God. Some religious people seem affronted by the suggestion that one can find evidence for the divine in the physical world. They make a firm distinction between matter and spirit, preferring to regard God as transcending the physical. There will always be gaps in our scientific understanding, it is affirmed, and that is where God comes in. However, it seems perverse to identify God exclusively with our areas of ignorance because He is surely either everywhere or nowhere. If He is everywhere, then one should expect science to reveal Him rather than conceal Him. The 'miracle' of matter should be no less astounding than the miracle of mind.

PART 3: INCORPORATING MIND INTO PHYSICS EXPLICITLY

So far our emphasis has been historical and mainly based on standard — if controversial — ideas of past and current science. This final section will look to the future and discuss whether there could be an extension of physics that incorporates mind explicitly. This means that our considerations will be much more speculative. We will also refer to findings in psychical research, a field which many scientists tend to reject out of hand.

Physics and Consciousness

One feature of the Universe which is noticeably absent in the unification table of Figure 1 is consciousness. Physicists have long between uncomfortable with attempts to incorporate this into their description of the world because consciousness is intrinsically private (1st person), whereas physics deals with things which are public (3rd person). Most physicists therefore assume that their focus of study should be the objective world, with the subjective element to be banished as much as possible. On the other hand, one might be sceptical of physicists' claim to be close to a 'Theory of Everything', when such a conspicuous aspect of the world is neglected.

Certainly physics in its *classical* form cannot incorporate consciousness. This was appreciated more than a century ago by William James, who stressed the incompatibility between the localised features of mechanism and the unity of conscious experience. Indeed some scientists - even psychologists - have tried to abolish reference to consciousness altogether. According to the behaviourist John Watson:

> The time seems to have come when psychology must discard
> all reference to consciousness; when it need no longer delude
> itself into thinking that it is making mental states the object of
> observation.

Although attempts by behaviourists to extend mechanism to the mind are now unpopular with psychologists, a mechanistic outlook still persists among many physicists and this probably contributes to their discomfort with consciousness. Daniel Dennett is even more forthright:

> Consciousness appears to be the last bastion of occult properties, epiphenomena and immeasurable subjective states — in
> short, the one area of mind best left to philosophers, who are
> welcome to it. Let them make fools of themselves trying to
> corral the quicksilver of phenomenology into a respectable
> theory.

However, other scientists are equally uncomfortable with attempts to formulate a picture of the world which makes no reference to mind (i.e. which neglects essentially half our experience). Thus the linguist Noam Chomsky declares: 'Physics must expand to explain mental experiences,' and Roger Penrose predicts: 'W e need a revolution in physics on the scale of quantum theory and relativity before we can understand mind.'

Of course, the classical picture of physics has now been replaced by a more holistic one and there are some indications that the new physics *can* include consciousness. In particular, quantum theory shows that, at its foundation, matter does not behave like a machine at all. On a microscopic scale at least, it introduces a new level of randomness and acausality into the world. It also brings the observer into physics in an essential way, so that John Wheeler can declare that: ' Mind and Universe are complementary.

Some people have even proposed that consciousness may be involved in the collapse of the wave function. Thus Louis de Broglie's studies of quantum phenomena convinced him that:

> The structure of the material Universe has something in common with the laws that govern the workings of the human mind

and Bernard d'Espagnat claims:

> The doctrine that the world is made up of objects whose existence is independent of human consciousness turns out to be in conflict with quantum mechanics and with facts established by experiments.

However, not all physicists accept these arguments. Many assume that quantum mechanical weirdness does not propagate up to macroscopic scales and simply cancels out for biological systems. Thus John Hopfield claims

> Contrary to the expectations of a long history of ill-prepared physicists approaching biology, there is absolutely no indication that quantum mechanics plays a significant role in biology

and Murray Gell-Mann asserts that 'no vital forces are needed for biology or self- awareness'.

There is thus the cherished hope that mechanism will survive at the level of the brain itself. However, according to Jeffrey Satinover, advances in neuroscience cast doubt on this hope. He argues that quantum effects play a crucial role in allowing randomness to arise in biological systems, so that they are no longer deterministic, with large-scale quantum effects being captured and amplified by the brain.

While this issue is clearly controversial, it is undeniable that quantum theory completely demolishes our normal concepts of physical reality, so it not surprising that some physicists have seen in its weirdness some hint of the mystical. It is certainly conceivable that some future paradigm of physics may incorporate mind, although it will probably go beyond quantum theory itself. In order to see what form that paradigm may take, we must stray into still more controversial domains!

Psychical Research

Psychical research studies those aspects of Man, real or supposed, which cannot be explained in terms of the current scientific paradigm. In particular, it examines phenomena such as telepathy (mind-to-mind communication), clairvoyance (direct perception of the physical world without the usual senses), precognition (foreseeing the future) and psychokinesis (the direct influence of mind over matter). Its findings also impinge on issues such as the survival of consciousness after death. If psychical phenomena are real, they necessarily involve an interaction between consciousness and the physical world. Therefore any theory of psi must amalgamate the mental and physical in some way and input from psychical research will be crucial in providing a 'bridge' between these worlds.

It seems likely that this bridge must involve a new paradigm, rather than just tinkering with the current paradigm, and that it must extend physics so as to incorporate consciousness explicitly. Furthermore, as discussed below, the nature of many psychic phenomena suggests that a crucial ingredient of the new paradigm must be some form of higher dimensional communal space, which transcends the usual sort of space and time. This space cannot the same as physical space but it must subtly interact with it.

• *Apparitions*

Many people claim to see apparitions occasionally. Although the images appear to be 'outside', they are normally interpreted as hallucinations created by the mind. On the other hand, some aspects of apparitions suggest that they really exist in some more objective sense. For example, they are sometimes seen by more than one person at a time or by different people at different times (as in the classical ghost case). There are even 'collective' cases where the apparition appears to be viewed from different perspectives, as though in the same space as the observers. However, the fact that apparitions rarely (if ever) leave any physical trace suggests that they do not exist in ordinary physical space.

- *Out-of-body experiences*

In an OBE, the point of consciousness appears to move away from the location of the physical body, sometimes even being associated with another 'astral' body. One feels 'awake' and seems to be moving around in some sort of space but it is subtly different from physical space. A sceptic would argue that the space encountered in an OBE is just a mental construct, with no relationship to the world encountered in the normal waking state. On other hand, there are occasions when one's consciousness appears to acquire veridical information about the physical world or even cause events there. Also at some stage one may encounter 'higher' planes, which are not related to the physical world at all.

- *Near-death experiences*

In an NDE consciousness may also move around in a quasi-physical space, just like the one encountered in an OBE. However, various other experiences are involved, such as the 'tunnel' effect, encounters with the 'light' and deceased love ones, memory playbacks and crossing a 'bridge'. The conformity of these experiences across a variety of different cultures might suggest that NDEs provide access to some higher form of reality, although sceptics might prefer to attribute this conformity to the similarity of the brain states of dying people.

- *Dreams*

All percepts would *appear* to exist in some form of space. For example, ordinary dreams - even those with no psychic content - seem to take place in a space which in many ways resembles everyday physical space. Although one usually assumes that this space is private, it can seem real and it provides the arena of many psychic experiences.

Combining Physical and Mental Space

In this final section, I would like to put the above remarks about psychical research into a more general context by discussing the relationship between matter and mind. All of us inhabit two worlds: there is the outer material world, which is studied by science, and the inner mental world, which we access while dreaming or in certain altered states of consciousness. There are similarities between these worlds: both seem to involve some form of space, so the description 'inner space' and outer space' is sometimes used in this context, and both involve the experience of time. However, there is an important difference between them: most scientists would claim the material world is 'real', in the sense that it can be accessed by everybody, whereas the mental world is 'imaginary'. More precisely, percepts associated with the physical world are assumed to be representations of an external reality, whereas those associated with the mental world are not.

Psychic phenomena, if real, would appear to refute this simple distinction. For example, the existence of telepathy suggests that some mental experiences can be shared and therefore possess some of the same attributes of externality as physical objects. Thus, if I visualise an image in my mind and somebody else 'sees' it, perhaps it really does exist 'somewhere', although presumably not in physical space. This links up with the above suggestion that many psychic phenomena require the existence of some form of non-physical space. If clairvoyance occurs, this suggests that the mind can directly apprehend the outside world (in the sense that it exists in the mind *a priori*). The distinction between internal world and external world thus becomes blurred and it may no longer make sense to ascribe reality to one but not the other.

I will now give this notion a more formal basis by proposing a definition of 'reality' which is more general than usual — in that it need not pertain exclusively to physical objects — but which still conforms to the general idea that something is real if it exists in an external space to which other consciousnesses have access. Of course, people who dismiss psi at the outset and assume that mental objects cannot have reality in the same way as physical objects will regard this as pointless. However, my intention is to provide a framework in which the question of whether mental percepts are real can at least be discussed.

If one were to ask a philosopher of the nineteenth century in what sense the physical world is real, he might have replied as follows: there exists a 3-dimensional space in which are localised both the sensors through which we observe the world and the physical objects themselves. Each observer has only partial information about that space because of the limitations of his sensory system. (For example, his eyes will provide him with a projection of the space which is essentially 2-dimensional.) However, the crucial point is that, given his location and the direction in which he is looking, one can always predict how he *ought* to see it. The fact that one can find a 3-dimensional configuration which predicts a set of 2-dimensional projections concordant with those which are actually presented to the different observers is what is meant by stating that the physical world is real. One may say that the physical world is a 3-dimensional *structure* which consistently reconciles how everybody within that world perceives it.

A modern-day philosopher would give a somewhat more sophisticated answer. Mindful of the implications of relativity theory, he would argue that the physical world is a 4-dimensional structure, with the objects and observers being represented by world-lines and the perceptual fields being 3-dimensional. However, the notion that the world is real because there exists a higher dimensional structure which reconciles our perceptions of it would be much the same. Indeed the prime message of relativity theory is that one can *only* reconcile how different observers perceive the world if it is 4-dimensional. Subsequent theoretical developments suggest that the reality structure which describes the physical world may require even more than four dimensions.

This discussion is fine so far as it goes, but it makes no reference to the other sort of sense-data which are presented to our consciousness: mental percepts with no physical counterparts. One therefore needs to extend the above approach to include the possibility that these may also be real (i.e. communal) in some sense. This can be done by assuming that the communal space has extra dimensions. The final reality structure is termed the 'Universal Structure' and the way in which it is constructed means that it has a hierarchical structure, each new dimension extending the range of phenomena to which can be attributed reality. It is natural to assume that the lowest member of the hierarchy corresponds to the 4-dimensional physical world of

Special Relativity, since our physical sensors do not directly access any higher dimensions. The Universal Structure may also be regarded as a sort of *information* space.

Since developments in physics also invoke a space of more than four dimensions, it is natural to identify these two spaces. For if our physical sensors only provide us with a 3-dimensional aspect of a Universe which in reality has many more dimensions, and if the physical objects themselves occupy only a limited part of that higher dimensional space, one is bound to ask whether anything else exists in this space. Since the only non-physical entities in the Universe which we experience are mental ones, and since the existence of paranormal phenomena (or even normal mental phenomena such as dreams and memories) would imply that percepts have to exist in some sort of space, it seems natural to identify this with Kaluza-Klein space. More precisely, we invoke a variant of the Kaluza-Klein paradigm, recently proposed by Randall and Sandrum, in which the physical Universe is regarded as 4-dimensional 'brane' embedded in a higher dimensional 'bulk'. However, it should be stressed that this sort of model for psi preceded the Randall-Sundrum proposal and indeed might be regarded as predicting it.

Although the idea of invoking higher dimensions to unifying matter and mind is not new, recent developments in physics have made this approach particular promising. Needless to say, this picture will not appeal to most of my physics colleagues, but then new paradigms always involve discomfort! Nor will it please some spiritual colleagues, since they may be uncomfortable with the notion that ineffable mystical experiences can be the subject of scientific modelling - or even rational discourse. Nevertheless, I suspect that such people will at least share my unhappiness with the brash dismissal of mystical experiences by so many scientists as mere illusions. It is only natural to try to extend the sort of theorising which science has applied so successfully to the material world to the mental world.

Conclusion

This paper has touched on all three of the topics promised by the title of this book — various aspects of science have been considered in Part 1 and Part 2, while consciousness and ultimate reality have been the focus of Part 3. The last part is much more speculative than the other ones and reflects my personal approach to the problem of amalgamating matter and mind. Nevertheless, I believe it is useful to have combined all three topics in this way. For if the sort of approach advocated here turns out to be correct, then mind is a fundamental feature of the cosmos in a much deeper sense than was discussed in Parts 1 and 2. If this turns out not to be the case, then the considerations of Parts 1 and 2 are still of interest in their own right, so even those sceptical about psychical research may find something of value in these parts.

Apart from references to divinity in Part 2 and mystical experiences in Part 3, I have avoided touching on theological issues. These are covered elsewhere in this volume. Suffice it to say that many of the arguments invoked here to prove the importance of mind have also been used to suggest the existence of God. They are not decisive arguments but they offer what John Polkinghorne describes as 'nudge factors'. In particular, the sort of paradigm proposed in Part 3 clearly impinges on the science-religion debate.

References

Barrow, J.D. (1990), *Theories of Everything*. Vintage Press.

Barrow, J.D. & Tipler, F.J. (1986), *The Anthropic Cosmological Principle*. Oxford University Press .

Carr, B.J. (1985), 'Mankind humbled in the progress of science', in *Taking the High Ground*. Saros Publications.

Carr, B.J. (1989), 'The Uroborus of science', in *The Saros Talks.* Saros Publ.

Carr, B.J. (1990), 'Science and the divine', in *The Saros Talks.* Saros Publ.

Carr, B.J. (1993), 'Life in the universe', in *Leaving Home.* Saros Publ.

Carr, B.J. (2000), 'The Anthropic Principle', in *Encyclopaedia of Astronomy and Astrophysics*, ed. P. Murdin *et al.* IOP Publishing & Macmillan.

Carr, B.J. (2004), 'Worlds apart: Can psychical research bridge the gulf between matter and mind?', *Proc. Society for Psychical Research*, in press.

Carr, B.J. & Rees, M.J. (1979), 'The anthropic principle and the structure of the physical world', *Nature* 278, p. 605.

Carter, B. (1974), 'Large number coincidences and the anthropic principle in cosmology', in *Confrontation of Cosmological Theories with Observational Data*, p. 291, ed. M.S.Longair. Dordrecht: Reidel.

Davies, P. (1988), *The Cosmic Blueprint*. New York: Simon & Schuster.

Dawkins, R. (1988), *The Blind Watchmaker*. Harmondsworth: Penguin.

de Broglie, L. (1963), cited in *The New World of Physics*, p. 143, A. March and I.M. Freeman. New York: Vintage Books.

Dennett, D.C. (1978), 'Towards a cognitive theory of consciousness', in *Brainstorms: Philosophical Essays in Mind and Psychology*. Montgomery: Bradford Books.

D'Espagnat, B. (1983), *In Search of Reality*. New York: Springer-Verlag.

Dicke, R.H. (1961), 'Dirac's cosmology and Mach's principle', *Nature*, 192, p. 440.

Dirac, P. (1985), cited in *Sympathetic Vibrations*, p. 225, K.C. Cole. New York: Bantam.

Dyson, F. (1979), 'Time without end: Physics and biology in an open universe',*Reviews of Modern Physics*, 51, p. 447.

Gell-Mann, M. (1995), cited in *The Third Culture*, p. 255, J.Brockman. New York: Vintage Books.

Greene, B. (2000), *The Elegant Universe*. Vintage Press.

Hawking, S.W. (2001), *The Universe in a Nutshell*. Bantam Press.

Hopfield, J. (1991), in *Evolutionary Trends in Physical Sciences*, ed. M. Suzuki and R. Kubo. Berlin: Springer.

James, W. (1950), *The Principles of Psychology*, reprint of 1890 text. New York: Dover.

Jeans, J. (1931), *The Mysterious Universe*. Cambridge University Press.

Kolb, E.W. (1996), *Blind Watchers of the Sky*. Vintage Press.

Kuhn, T.S. (1970), *The Structure of Scientific Revolutions*. Chicago University Press.

Leslie, J. (1989), *Universes*. London: Routledge.

Linde, A. (1990), *Particle Physics and Inflationary Cosmology*. Harwood Press.

Newton, I. (1692), Second letter to Richard Bentley.

Pagels, H.R. (1992), *Perfect Symmetry*, p. 359. New York: Simon & Schuster.

Penrose, R. (1997), *The Large, the Small and the Human Mind*. Cambridge University Press.

Pippard, A.B. (1988), 'The invincible ignorance of science', *Contemporary Physics.*, 29, p. 393.

Polkinghorne, J. (1994), *The Faith of a Physicist*. Princeton University Press.

Randall, L. & Sundrum, R. (1999), 'An alternative to compactification', *Physical Review Letters*, 83, p. 4690.

Reeves, H. (1991), *The Hour of our Delight*. Freeman.

Satinover, J. (2001), *The Quantum Brain*. Wiley.

Smolin, L. (1997), *The Life of the Cosmos*. Weidenfeld & Nicolson.

Smythies, J.R. (1994), *The Walls of Plato's Cave*. Avebury Press.

Weinberg, S. (1977), *The First Three Minutes*, p. 154. New York: Basic Books.

Wertheim, M. (1999), *The Pearly Gates of Cyberspace*. Virago.

Wheeler, J. (1977), 'Genesis and observership', in *Foundational Problems in the Special Sciences*, p. 1, ed. R. Butts & J. Hintikka. Reidel.

Chris Clarke

Quantum Mechanics, Consciousness & the Self

Introduction

The phrase 'Ultimate Reality' is huge and sweeping, evoking something ungraspable, perhaps too far away from our lives to be of any real relevance. So I would like to begin by bringing it down to earth as the more homely question, 'what is the world really like?' Phrased in that way, it starts to feel more close at hand. Is the world, for example, a meaningless machine constructed of identical particles? Is it a manifestation of a loving Spirit? These, and innumerable other possibilities are the germs of quite different stories that we might tell about the nature of the world. And the story that society tells to itself about the world really matters, because it affects, or it expresses and reinforces, the values of that society. These values may, in turn, affect or reinforce its power structures — or in times of change may undermine them. The whole social fabric within which we live is inseparable from the question, what is the world really like?

The values of the world that most impinge on me are the values of modern capitalism: that the highest good is the gratification of the desires of isolated individuals, support by a planet which only has value as a (bottomless) resource for human beings. Its power structures, matching these values, are those of the multinational corporations which view the individual purely as a consumer and society purely as a vehicle for encouraging consumption through the desire to conform. The story that reinforces these values is the mechanistic story started by Newton's successors (such as Laplace and Lemaitre) of a meaningless universe of isolated atoms. Beginning in physics, this story is now continued in the molecular biology that has taken over from physics as the dominant strand of science. On this story, we have no true connections with each other or with the other beings on the planet because we are made of disconnected atoms. Our role is that of mechanistic consumers because the material of which we are made has no ultimate freedom. Meaning and purpose are comforting illusions with no basis in reality.

Though many of us, perhaps the majority, do not actually believe — or even consciously know about — this story of the world, it is enshrined in our educational curricula, in our health care systems (dominated by mechanistic biology), in our environmental actions, in our economics. Since physics gave rise to this story, so now physics has a vital role in holding the culture-bearers of our society to their current destructive course.

How has it come about, historically, that a particular version of physics is now laying claim to a description of ultimate reality? I'd like to give my own (grossly simplified) version[1] of this. I think of the Western stories of what the world is like in terms of the categories of **stuff, change** and **layers.**

What sort of stuff constitutes the universe? The obvious answer might be just a list of 'things' — trees, birds, rain, demons, angels, stars. . . . More subtle thought might lead us to choose something more unified and basic, such as numbers (Pythagoras) or The Void (Anaximander), or Water (Thales), or some combination of basic elements (Empedocles). The stuff might be continuous or it might come in elementary particles, atoms or seeds (Democritus). Descartes held that there were two basic sorts of stuff, matter (*res extensa,* stuff with extension) and mind (*res cogitans,* thinking stuff), with quite different ways of behaving (a view often called *substance dualism*). Modern physics, in a sort of return to Pythagoras, sometimes regards the stuff of the world as being abstract and mathematical.

---

[1]   I draw this mainly from Cleve (1965).

The world seems to change. We might think — perhaps in the light of mystical experience — that change is some sort of illusion (Parmenides), or a secondary aspect of the world. Or it might be absolutely fundamental (Heraclitus). Change might be the result of changes in form of a changeless underlying stuff, or it might be a process in which some things stopped existing and new ones started existing. Does an acorn turn *into* an oak tree, or does it die to be replaced by an oak? Change might be governed by rules, as happens with the laws of modern physics, or it might be, at least partly, chaotic and anarchic as was supposed by Epicurus and Lucretius.

Finally, there might be a distinction between what we are immediately aware of and one or more other 'layers' of the world which are more real or fundamental than the one we are aware of. This was a view made famous by Plato, who stressed the greater reality of the world of forms and ideas, and which evolved into the dualism of Descartes. The cosmologist George Ellis (Murphy & Ellis, 1996) has revived and extended the idea in the form of a multiplicity of 'worlds of discourse' corresponding to the physical, the human, and so on, whose relations may be very complex. The dissident physicist David Bohm spoke in terms of an explicate layer that was unfolded from a more basic implicate layer.

These various attitudes to stuff, change and layers have been interweaving for 2700 years. Crucially for understanding our present position, they crystallised in the West during the eighteenth to early twentieth century, before then starting to re-melt and enter a new period of flux today. The crystallisation was initiated by the mind/matter dualism of Descartes, and in particular by his view that mind made up the human soul, which was a single indivisible entity having no place in physical space, equally present to every part of the body (though the part of the body most sensitive to its presence was the pineal gland). The endorsement of this view, first by the church and then by the emerging scientific establishment, installed it as the accepted reality for Western thinking. At the same time, the growth of the idea that the material side of this dualistic stuff consisted of tiny particles installed *atomism* as the accepted way of thinking about matter.

Perhaps surprisingly, it was this dualistic split that was responsible for the mechanistic story within which we now live.

What happened next was a process of exploration of what functions were exercised by mind (or spirit, as some called it), and what by matter. As science developed, more and more theories emerged as to how matter could in fact exercise many of the faculties that had previously been attributed to mind/spirit: the origin of computers clinched the argument, for many people, that mind was unnecessary and matter could do everything. The argument for physics being Ultimate Reality might run as follows: 'We are finding physical explanations for more and more phenomena, and the need for anything like spirit as an explanation is dwindling. Though the explanations for phenomena are often couched in terms of sciences such as astronomy, biology and geology, these sciences all depend on physics for the description of the basic matter of which everything is composed, and in principle all these sciences could ultimately be reduced to physics.'

I want to maintain at the outset that an argument like this, for what might be called physicalism, cannot actually be justified. It starts from Descartes' absolute division of the world into the separate substances of mind and spirit, for which there is no evidence. The idea of 'phenomena' presupposes the same division, in which our perceptions are mere mental appearances produced by a separate physical reality. The 'in principle' reduction of other sciences to physics is an act of faith that is impossible to prove. The concepts of 'physical' and 'matter' have repeatedly changed through history as attitudes to stuff and change alter, and are now so fluid that this argument cannot rest on them. The whole argument for the modern story is a flimsy construction resting on one basic intellectual move, the adoption of substance dualism.

My aim in this chapter is to question the nature of this division in the light of modern science. We need to start again, in the expectation that physics will be just one contributor to a growing understanding that draws on all facets of our knowing and being. I shall describe here the new way in which physics now contributes to this understanding, while needing ideas about meaning, consciousness and creativity that lie outside its normal domain. The argument will have two main steps. First, I will explain the collapse of the old idea of philosophical atomism: the idea that we need to look to the small scale level of particles in order to find reality. Second, I will describe how, in the light of this, physics stands in need of new ideas, and what shape these ideas need to take.

Quantum Mechanics and the Death of Philosophical Atomism

The birth of quantum mechanics

The so called 'Classical Mechanics', which emerged largely as a result of the work of Galileo and Newton, had a clear picture of absolute physical reality, an ultimate specification of what was the case for the physical universe at any given time: namely, the exact positions and velocities of every particle in the universe at that time. No one could know this, of course, but the fact that this 'existed' in some hypothetical sense said something about the nature of physical reality. 'Stuff' was particles, and it had a decidedly mathematical flavour, being entirely specified by mathematical quantities. Change was lawlike and governed by mathematical laws. And by the eighteenth century the world essentially had only one layer (heaven being so far removed that it could not be considered part of the same world). In classical mechanics, every other property of the physical universe could be defined precisely in terms of the ultimate specification of the positions and velocities of particles. Ultimate Reality was thus to be found at the level of the very small, in the atoms (in the sense of 'whatever particles were ultimate') and the laws that governed them. The first step in my argument is to explain how this idea has been eroded by the quantum mechanics that has replaced classical mechanics. This will in turn lead on to the idea that physics cannot provide any sort of ultimate reality at all, although it can give some very strong clues about the direction in which we need to go in order to get closer to 'ultimate reality' (if that notion makes at least some limited sense).

I must, however begin with quantum mechanics (or 'quantum theory' — I am not here making a distinction). It is a remarkably diverse discipline. On the shelves of this section of a science library one can find books relating to philosophy, cosmology, particle physics, optics, electronic engineering and much else. Even if one is familiar with the subject, it is almost impossible to tell what quantum mechanics 'is.' Rather than trying to define it, I shall take a roughly historical approach leading to some central ideas that are important for the theme of this book.

Quantum mechanics started in 1900 when Max Planck tried to understand the appearance of hot bodies. We are familiar with how heated metal first glows a dull red colour, then, as the temperature is increased, a brighter red, then moving to white and blue. What determines this? What was expected to be a simple problem turned out to require a radical solution: that light was not continuous, but made up of packets which Planck called *quanta* (the plural of *quantum*, which is the Latin etymological root of the word 'quantity'.) The quanta of blue light had more energy than the quanta of red light and the colour of the light emitted by a hot body depended on a balance between the energy of vibration of its atoms due to heat and the amount of energy in the quanta of light of different colours. Quanta of light are now called *photons*; we will meet them again later.

That historical story explains the name, but I'm sorry to say that it is otherwise almost completely irrelevant. Although the early history of the subject was dominated by the idea that energy sometimes (not always) came in packets, after a while it became clear that this was not really a fundamental point. As fundamentals, I would choose two later discoveries: *complementarity* and *non-locality*.

The implications of complementarity

'Complementarity' is a complete misnomer: the subject should be called 'incompatibility', but we are stuck with the former name for historical reasons. Think first about the picture given by classical mechanics in which all physical quantities can be derived from an ultimate physical reality consisting of the specification of the position and velocities of all the particles in the universe. Complementarity/incompatibility says, in contrast, that in quantum mechanics *there does not exist* any consistent specification of the values of all the properties of *any* physical system (apart from the most trivial ones).[2] We can chose some subset of its properties and specify their values, but there will come a point when there remain properties that cannot be specified without being inconsistent with some of the ones we have already specified.

This seems a radically different state of affairs from classical mechanics. In fact, it throws into the air again all the previously

---

[2]   This formulation is the Kochen-Specker theorem (Kochen & Specker, 1967).

established ideas about stuff, change and layers and undermines the idea of philosophical atomism. What is now the stuff of the universe, if one cannot, even in principle, specify all its properties? Is there a more fundamental layer (the quantum state? the implicate order?) where things are specified, but inaccessible to us? If properties are not specifiable, then must the laws of change become partly anarchic — or might they still hold precisely in a more fundamental layer? (Both these are probably the case in quantum mechanics.) Although the case is not totally clear-cut, the development of quantum theory strongly suggests that *the physical universe has no ultimate reality at the microscopic level* in the way that the Newtonian physical universe has.

In fact, we can be more specific than this about what is going on. We certainly can carry out observations to determine particular microscopic properties of the quantum realm, but some of these observations are incompatible with each other. The operation of the observer — that is to say, an operation in the large-scale world defining a particular context — determines which properties of the small-scale world are going to have values. Physical reality (at least, in so far as we can know it, or in so far as it is manifest to us) is not something that is given at the small scale, and it then trickles up to the large scale. Rather, this sort of reality arises from the interplay between the large-scale context and the small-scale system.

This will make a crucial difference to the story, so we need to proceed carefully. Many physicists would contest the idea that quantum mechanics undermines the classical picture, on the grounds that quantum mechanics is only to do with the very small which we cannot perceive anyway, and so it only gives a minor correction to the Newtonian picture, leaving it largely intact for most purposes. This argument has in fact an element of validity, so it is important to understand its basis. This takes us back to Max Planck. The size of his packets of light was determined by a constant, soon named Planck's Constant, whose value is very small. It emerged that the real significance of this constant is that it specifies the extent to which the different properties that we might want to specify in the universe fail to fit together consistently. To be precise, for every quantity describing a physical system, there exists at least one complementary quantity (hence the name complementarity) such that neither can be specified at the same time. The best that can be done is to specify both approximately, and the product of the uncertainty in each is at least the size of Planck's constant. (This is the Heisenberg uncertainty relation.) The point is that because this constant is very small, at large dimensions — say, at the size of large biological molecules — all physical quantities can *almost* exactly be specified at the same time. This means that a lot of physics can continue as if classical mechanics were still true.

This does not mean, however, that quantum effects are confined to the very small. This is one of those cases often occurring in physics where a small constant can have a large effect. A more familiar example is that of viscosity, which causes turbulence. If one compares flows over the same dimensions and with the same speeds in a fluid with a large viscosity, such as treacle, and in one with a small viscosity, like water, it is the one with the *smaller* viscosity that shows the *larger* turbulence. Introducing a change that is numerically very small can produce huge changes in the overall behaviour of the system. It is the same with Planck's constant and quantum theory.

The implications of non-locality

The second of the two fundamental discoveries of quantum theory is non-locality. 'Local' means to do with place, or restricted in space, as opposed to 'non-local'. There are two concepts in physics that encapsulate this distinction: fields and states. A *field* expresses purely local properties.[3] At each place and each time a field has a definite value. For example, the electric field at a given point specifies the direction and intensity of electric forces there; and the gravitational field specifies the same for gravitational forces. Fields change in time according to laws which are also local, in the sense that the way the value of a field at a given point of space changes is determined by the values of the field only at neighbouring points, and not by values at a distance.

The *quantum state* (usually abbreviated simply to '*state*') of a system is, by contrast, highly non-local. It is a rather more problematic concept, because it can be expressed in many different ways in physics and thought about in different ways conceptually. If we think of it in relation to the past, it expresses the cumulated effect of the history of the system at a given stage; if we think of it in relation to the future, it expresses the probabilities

---

[3]   A confusing issue here is that in New Age 'Physics' the word 'field' is used to evoke something non-local!

for all possible properties of the system that might be observed or manifested next. The state depends on the entire system, not on local parts, and so it is fundamentally non-local. It is normally thought of as the state at a given time; but the notion of 'time' becomes questionable in relativity theory and this is probably not an accurate way of thinking about it.

While non-locality is always an essential aspect of the state, there are particular physics experiments that reveal this explicitly. The most famous of these, performed by Alain Aspect, involves producing two photons (quanta of light) from a single 'mother' photon, letting them separate by a considerable distance, and then simultaneously examining their properties. What is observed is a correlation between the properties that the two manifest. Moreover, the particular sorts of correlations observed are such that they could not be obtained by any properties of the photons that are local to each one. This (known as Bell's theorem) depends hardly at all on any physical assumptions, but arises just from the logical consequence of the particular sorts of correlation observed. It is as if the photons are responding jointly in a co-ordinated fashion to the observations, even though there is no physical possibility of any communication between them.

The explanation given to this by the formalism of quantum theory is that the two photons share a single quantum state (which, it will be recalled, is non-local). The photons cannot be thought of as separate systems, even though they appear in two different places. This is a concept-bending idea. We think of 'things' in terms of space: if two lumps of stuff are separated in space, then they are different 'things'. But in quantum theory the 'thing' that is a quantum state ignores space entirely, embracing appearances that we see as separated as if the separation did not exist. There is a useful terminology for expressing this. When the quantum state of what appear to be two systems cannot in fact be separated into two quantum states, one for each apparent system, we say that the systems (or their states) are *entangled*.

This second aspect of quantum mechanics further undermines philosophical atomism, because it indicates that even when we get down to the level of basic particles, these particles are not independent, as supposed in atomism, but integrated into larger systems by an underlying web of non-local entanglements.

Pointers from Quantum Theory
to Human Experience

So far I have described the impact of quantum theory in negative terms, showing how it has undermined the old certainties about the nature of physical reality. This is important, because if the old certainties go, then the way is opened up for a new story. But for many thinkers, the importance of quantum theory lies even more in what it might say positively about the nature of human beings (and, indeed, organisms generally) and their place in the world. This would enable quantum theory not only to clear the way for a new story, but to start telling that story together with other areas of enquiry. I want to outline these possibilities here, expanding on some in what follows. First, however, I need to offer a word of warning. All these 'pointers' from quantum theory depend on the extent to which particular aspects of quantum theory extend from the microscopic world, where they have been tested, to the large-scale world, where on the whole they have not. It is certainly the case that some properties (notably, indeterminism, a relaxation in the rule of rigid laws) do extend in this way. But whether all properties of quantum theory extend depends on the outcome of a debate on a phenomenon called *decoherence,* to which I will return later.

Indeterminism and human freedom

In classical mechanics the laws of change are *deterministic:* they specify exactly what is going to happen next, and there is no room for lee-way. In classical mechanics fate is inexorable. In quantum mechanics the laws of change are *indeterministic:* instead of saying definitely what will happen, they say that some things are more likely than others — they give probabilities for different outcomes, no more. This is one property of quantum theory which extends without any doubt to the large-scale world, because we know innumerable mechanisms, both natural and artificial, which can readily amplify an uncertainty at the very small scale to one on a very large scale. Turbulence, which I mentioned in the last section, is a case in point. Here a microscopic change in the ways atoms move in one part of the river can grow so as to alter completely the shape of a wave further down the river.

As human beings we feel that our own actions are not determined by inexorable laws. This is part of what we might mean when we say that we are free. So on the face of it this property of quantum theory seems important for understanding what it is to be human. This could, however, be misleading.

First, we don't really need quantum theory to produce *de facto* indeterminism. Just as the waves in a stream can be altered by quantum mechanics, so they can be altered by the random motions of molecules that are happening all the time in classical mechanics. Admittedly, in the latter case the laws are 'really' deterministic and the motions of atoms are 'random' only in the sense that there is no way in which we could know what they are. There is a difference in principle here with quantum mechanics, where the uncertainty arises from the basic logical structure of the underlying 'stuff'. But it can certainly be contested whether this difference is real or only formal.

Second, human freedom does not merely depend on a chink in the determinism of physical law: we must also have the capacity to *act* within that chink so as to affect the flow of events. We might know that the motion of a few atoms near the top of the stream could create a deluge near its mouth, but nothing can tell us which atoms to move, when, and how far.

Third, we might be deluded as to the nature of our freedom, imagining we are free agents in events when we are either unconsciously driven or acting randomly, and/or ignoring those events where we might really exercise freedom. Indeed, the nature of free will is already a vexed philosophical question which needs careful analysis before we can decide just what it is that we hope to get from quantum theory.

For all these reasons, this aspect of quantum theory that is most certain is still contested when it comes to linking the theory to human experience.

Quantum logic and creativity

The incompatibility of physical properties implied by the Kochen-Specker theorem can be understood as saying that the way in which these properties fit together has a different formal structure from the way in which they fit together in classical mechanics. By 'formal structure' I mean what mathematicians call the *logic* of the propositions corresponding to these properties. At this stage I need some more terminology, relating to mathematical logic. A *proposition* is a statement that might (at least in the simplest cases) be true or false; such as 'this book weighs 500g' or 'all sycamore trees bear seeds with an attached wing.' By 'fitting together' propositions, I mean combining them with conjunctions such as 'and' or 'not' so as to form *compound propositions*. And in such cases the *logic* is the set of rules for determining the truth or falsity of compound propositions in terms of that of their components, and determining the equivalence or otherwise of various compound propositions.

To take an example derived from one in Chris Isham's talk in the project of this book: suppose I go into a restaurant and note from the menu that I can have sausages with either fried eggs or poached eggs. I accordingly order sausages with fried eggs, to be told that this is unfortunately unavailable; I therefore change to sausages with poached eggs, only to be told with great regret that this is also unavailable. Yet when I direct the waiter to what is written in the menu, he affirms that what it offers is nonetheless available. I am confounded, because my logic contains the rule

> if *A, B* and *C* are propositions, then the compound proposition 'A and (B or C)' is equivalent to '(A and B) or (A and C)'.

(This is called the *distributive law* of formal logic.) I was taking *A, B* and *C* to stand for 'sausages', 'fried eggs' and 'poached eggs', respectively, and supposing that 'available' worked like the logical condition 'true'. The waiter, on the other hand, seemed to be using some different logic, in which either this law did not hold, or 'available' did not obey the same rules as 'true' in formal logic. My suspicion was confirmed when I ordered 'sausages with either fried eggs or poached eggs' and was served with sausages and scrambled eggs, being assured by the waiter that 'fried or poached' obviously implied all other sorts of eggs, some of which were in fact available.

The waiter was in fact using quantum logic, a system in which all the usual laws hold except for the distributive laws. Its characteristic is that it is not possible to assign values of 'true' or 'false' in a consistent way to all the propositions (the Kochen-Specker theorem again). Instead one moves from one limited set of propositions to another, keeping the assignments consistent within each set, but with there being no universal concept of true or false. Now this way of thinking is actually what we do all the time. During a convivial evening in the pub our conversation wanders in ways that, in the clear light of the next morning, would appear to defy all laws of classical logic. But more importantly, it is also the logic of creativity — as when Picasso constructs a sculpture of a baboon using a toy car for the top of the head, which sits there defiantly challenging the viewers' logic by being both a car and a baboon's head. It is a logic I use when grappling with a scientific problem and bringing in all sorts of analogies with other problems that I have solved in the past. I would contend that it is actually in these flights of creativity where our real freedom lies, and not in the bare fact of indeterminism. We have the capacity to work between different *frameworks of meaning* (technically, Boolean sub-logics) which are inconsistent with each other, in order to create new ways of looking at, and acting in, the world.

Non-locality and the qualia of perception

We will see later that, according to one philosophical 'school', the characteristic of consciousness is that it is the 'view from inside' which carries with it qualitative aspects which cannot be translated into formal structures that are communicable to others as simple information. We cannot explain to a congenitally blind person what 'red' is like; only shared subjective experience, from the inside, can do this. This idea was introduced into the study of consciousness by David Chalmers (1995), based on the seminal paper of Nagel (1979). Where do these aspects of experience (called *qualia*) come from?

An attractive explanation can be found in terms of non-locality and entangled systems. Whenever two systems interact, they become entangled, though it is only in exceptional circumstance that this entanglement has consequences that are observable to physics. Thus when I interact with something by observing it, the part of my brain responsible for conscious awareness is entangled with an aspect of the thing observed. In other words, though they appear two distinct systems, they are in fact a single system. So, on this way of talking about it, an observed aspect becomes actually a part of my conscious self. This could explain how its qualitative properties become accessible to me, and how the content of conscious awareness is 'out there'. A great deal more needs to be said here (Clarke, 2002), because of the way in which purely logical processing is always intertwined with this primitive grasping of qualia. In the present context, the problem is that since everything is in fact entangled with everything else, is the entanglement with our brain such that objects of perception are specifically and identifiably entangled with their corresponding brain states? This involves understanding the extent to which the distinctively quantum mechanical property of non-locality extends to large-scale phenomena.

Top-down causation and the action of mind

In quantum mechanics, because of quantum logic, the framework of meaning within which a property manifests itself is determined by the wider context, which might be that of a human observer in a laboratory, or it might just be some physical process that creates a permanent record. This is the celebrated observer-dependence of quantum mechanics. It implies that the large-scale context has a causal influence on small-scale manifestation ('top-down' causation) as well as the small-scale effects building up to large-scale phenomena ('bottom-up' causation). I have already discussed the way in which this might operate within our brains to enable us to switch frameworks of meaning in creative thinking. But if our brains are entangled with the external world, then this top-down causation will automatically carry the external world with it. This particular sort of linking with the external world is well understood and operates, for example, in the theory of the way in which particle-detectors work in a laboratory. When a particle is detected by an array of detectors, the manifestation is consistent throughout all components of the laboratory. Although the focus of the event is a quantum mechanical (non-classical) microscopic system, the result is inevitably a consistent macroscopic phenomenon. What is controversial in the present discussion is the idea that the focus of such an event might be a comparatively large object like a system in the human brain.

This then gives us a mechanism whereby the indeterminism of quantum theory could actually be used by us to exercise our freedom in action. We do not have to know, either consciously or unconsciously, the physical processes that connect our brain states with the outside world. What happens is that these processes will automatically ensure that the outcome in the world is consistent with the framework of meaning that we create in our brain. It is important to stress that I am not talking here about 'direct mind action' of the sort sometimes considered in discussions of psychokinesis or healing. Here the meaning-making of my brain becomes linked into a greater meaning-making of all the organisms in the universe, those that we would normally regard as animate and those that we would normally regard as inanimate.

Pointers from Human Experience to an Understanding of Quantum Theory

The incompleteness of physics

I have argued that the advent of quantum theory entirely undermined the classical view that was previously an accepted foundation for the world, and that as a result the nature of the world has become an entirely open question. In particular, while contemporary physics offers many fascinating pointers to human experience, it cannot at present be seen as in itself a sufficient foundation for that experience — first, because these pointers still need to be translated into full theory, and second, because mainstream physics is itself incomplete and stands in need of enlargement or of input from outside physics. I want now to examine this second aspect.

The way we think about quantum theory depends on what attitude is taken to the categories of stuff, change and layers. In some approaches the stuff of the universe is the quantum state, that abstract, non-local entity that encodes the past and provides probabilities for the future. In these approaches, which I shall call *state-based*, the quantum state is thought of as a real thing that directly underlies what we experience. At the level of the state, change is governed by an equation, called Schrödinger's equation, which is deterministic. However, in order to avoid the occurrence of states which cannot possibly correspond to our experience, some extra dynamical principle is introduced which occasionally causes the state to change very rapidly, a phenomenon called the 'collapse of the state'. This introduces indeterminism. An example of a state-based theory is that of Roger Penrose. It requires the development of a unified theory of gravitation and quantum theory, one of the major challenges of physics today, and is thus incomplete until this is achieved.

A second approach derives its 'stuff' only from the larger scale phenomena of which we are directly aware as real, and regards the quantum state as no more than a shadowy device for expressing the way these phenomena are linked and evolve. I shall call these *macro-based* theories. In these the need for an additional factor is sometimes not made explicit. For example, the many-worlds approach supposes that the universe splits into branches when any 'observation' takes place, but usually only vague conditions are given for what exactly constitutes an observation. The devil is then in the (omitted) details.

Mention should also be made of the version of quantum theory proposed by David Bohm (see Bohm & Hiley, 1993). This is a 'layered' cosmology that separates the world into explicate and implicate levels, and in its later development it appears that Bohm envisaged many different levels of the implicate order. The stuff of the universe, in the simpler forms of the theory, consists of particles and fields, and change comes through a special set of equations that have a deterministic form, even though their result may not be deterministic. It is not clear whether or not this version is complete, even in this simplest form, because of technical issues concerning whether or not initial conditions uniquely determine the subsequent development of the universe.

Here I shall pursue one particular macro-based approach, namely the *histories interpretation of quantum theory*, which I think can lead to a very clear understanding of just what is required to extend quantum mechanics to a complete theory. Recall that the state, which is a basic object in the original form of quantum theory, can be thought of as encoding the effect of past history. In the histories interpretation this history is itself taken as the stuff of the theory and the state takes on a secondary role. A 'history' here is a sequence of assertions about the world — such as: 'first there is a cloud of gas; then it forms galaxies; then they produce stars; then at least one of them produces planets ...' (or in could be something more detailed, referring to laboratory physics).

Quantum mechanics is then interpreted as a procedure for calculating the probability of any given history. For example, if the probability of the above history works out to be ½, then we think that there was a 50:50 chance of the universe containing planets, and so the appearance of Earth was not that remarkable. A probability of 0 means that the proposed history is so unlikely that it is essentially impossible, while a probability of 1 means it is so certain that it is essentially inevitable. One can see from this that 'change' is of rather an anarchic form. Anything can happen, except that it is constrained by the interplay of probabilities.

I have been vague about what sorts of assertions might belong to a history, and this vagueness in fact conceals the incompleteness of the theory. Because, if we already have some idea of the sort of things that we might want to put into a history then we are already implicitly introducing into the picture some degree of specification of the universe. In the example I have just given about the formation of planets, for example, there are lots of hidden assumptions about the sorts of cosmological pictures that we think are roughly correct in the first place, and we are then invoking physics to decide between models that we have already selected in this way. What would happen if we dropped this pre-selection of what we already knew about the universe and included in our histories everything that corresponded to some sort of macro-scale quantum state? In that case we would open up far too many possibilities. First, we would be introducing so many possible histories that their probabilities would add up to more than 1. (Imagine rolling a die and being told as it rolled that every face had a probability of ½ of turning up.) Second, almost all quantum states correspond to situations that simply are not observed. Namely, there are vast numbers of states which arise from a purely mathematical operation of taking a superposition (an sort of average, discussed later) of two acceptable physical states, giving something that is well defined mathematically but has no real interpretation. For example, one can define a state that is a superposition of my being in my bedroom and my bathroom. This does not mean that I am in the corridor in between. If it were to mean anything, it would have to mean that to a certain extent I am in the bedroom and nowhere else, and to a certain extent I am in the bathroom and nowhere else — and this makes no sense in terms of our ordinary experience. We need somehow to alter or complement the physics so that it agrees with what we are actually aware of. (I return to this example in the section on decoherence, below.)

Something therefore needs to be done to pare down the events that can appear in a history. In the past, what has been done has been simply to *impose* the condition that the probabilities add up to 1. A selection of such histories is called a *consistent* set of histories. The trouble with this is, firstly, this is arbitrary, putting in by hand something that one cannot explain through physical theory. And secondly, even this doesn't work because it still allows the occurrence of histories containing the sort of impossible states that I have just described. We have to do better than this, and the clue comes from a requirement that I mentioned above, namely we need a physics that '*agrees with what we are actually aware of*'. In order to achieve this, surely we have to include in our theory something about the nature of awareness? In other words, a complete account of the world as we know it has to include an account of our knowing, our awareness.[4] Thus we find, not only that quantum theory offers pointers to human experience, but also that it stands in need of human experience in order to be a fully complete theory.

Consciousness

There is a long history of using consciousness as a means of completing quantum theory, running from Wigner to Penrose. Most of these ideas fall within the category of state-based theories, with consciousness playing the role of a factor influencing the evolution of the state. Penrose's theory is an exception, in that it is gravitation that influences the state, and the entire process is manifested as consciousness. The other theories in this class suffer from a problem common to most forms of interactionist dualism, where consciousness is regarded as a separate layer of reality from the physical: what exactly is it that consciousness is supposed to do, that material entities cannot do, and how exactly does an immaterial consciousness interact with a material universe? In recent years, however, the need has emerged for a new ingredient, which Zeh (perhaps misleadingly) has named a *theory of mind*. By this he does not mean a separate layer of reality which interacts with the physical; rather, it means a theory about which sorts of things *in the physical world* correspond to what we call 'mind', in the sense of being associated with awareness. It is these things that are then singled out in our physics as the place where we apply the criterion that the theory should agree with what we are aware of. Our physics must be such that minds turn

---

[4]   An argument can be made here that what matters for quantum theory is not the fact of being aware, but the fact of a record being made of the external world. Against this, it can be argued that (a) in the absence of intelligent beings to interpret the 'record' it is not clear what does and what does not constitute a record; (b) it seems strange to exclude from consideration things of which we are momentarily aware, but then forget, leaving no record; (c) philosophically it is awareness that is fundamental, for everything that we conceive and theorise about has its roots in our immediate awareness of the world around us.

out to have experiences that are (in some appropriate way) like ours.

When we talk about awareness, or experience, we need to clarify the sense of this. We can either think of awareness 'from the inside', from the point of view of the mentally endowed organism having the experience (the first-person perspective), or 'from the outside', from the point of view of an onlooker noting what changes in the behaviour of the organism having the experience (the third-person perspective). Here I will be concentrating on the former perspective when I talk about consciousness, and I adopt the view of Max Velmans in his book *Understanding Consciousness*, where he makes gives the following account of what he means by the word:

> We have knowledge of what it is like to be conscious (when we are awake) as opposed to not being conscious (when in dreamless sleep). We also understand what it is like to be conscious *of* something (when awake or dreaming) as opposed to not being conscious of that thing.
>
> This everyday understanding provides a simple place to start. A person, or other entity, is conscious if they experience *something*; conversely, if a person or entity experiences nothing, they are not conscious. Elaborating slightly, we can say that when consciousness is present, phenomenal content is present. Conversely, when *phenomenal content* is absent, consciousness is absent.
>
> ... The 'contents of consciousness' encompass all that we are conscious of, are aware of, or experience. These include not only experiences that we commonly associate with ourselves, such as thoughts, feelings, images, dreams, body sensations, and so on, but also the experienced three-dimensional world (the phenomenal world) beyond the body surface.

It is easy to be distracted from this basic idea of first-person experience. It is an idea that is so simple, so immediate, that we cannot see it; as in the saying that the fish is unaware of the water. Consciousness, in this sense, is not self-consciousness, nor noticing things, nor problem solving, nor talking to oneself. . . . it is simple awareness.

To return to physics, then: recall that we were inquiring whether a theory of mind, a theory of what things were aware, could make modern physics complete by using mind (including the mind of human beings, of course) to select the possible contents of histories. Using this definition, we can say that the theory of mind needs to be a *theory of consciousness*.

Can a theory of consciousness be formulated in terms of physics as it is — or, if not, how far do we have to go beyond physics? To clarify ideas here, it is helpful to look at one of the more successful examples of a theory of mind, due to Donald (1995). His starting point is the fact that our awareness is associated with our brain, which seems to have the characteristics of a sort of switching mechanism, and so his theory of mind becomes a theory of particular sorts of switching mechanisms. But for such a theory, a sophisticated telephone exchange could also qualify as having a mind, and why should a telephone exchange be aware of anything? Generalising this idea, *whatever* physical mechanism we describe, and however complex the things it does, we can still suppose that it performs these functions entirely without consciousness. Consciousness, in the sense used here, is by definition a first-person perspective idea, and the specification of a physical structure is a third-person perspective idea. One cannot derive the former from the latter. This argument is the basis of the position of David Chalmers, which I described earlier, that no theory of consciousness can be constructed on the basis of specification of function or on physical description.

As Chalmers shows, while one cannot proceed from a third-person account (such as physics) to a first-person account (consciousness), it is, however, possible to proceed in the opposite direction: if consciousness is taken as a primitive element in the theory (and all theories must have some primitive elements) then we can define third-person, objective concepts in terms of first-person concepts by extracting those elements of first-person views that people share in common. By reversing the usual order of argument, a way, that is not fundamentally flawed at the outset, is opened up for determining which things are conscious.

A theory of consciousness introduced in this way cannot be regarded as physics in the usual sense because it deals essentially with first-person, subjective experience. I take this as a strong, though not coercive, argument for the idea that physics is in itself incomplete and must be augmented by a subjective account of consciousness. The importance of this argument is that it tells us how the theory of consciousness then fits in with physics: it selects from the array of possibilities that might appear in the histories described by physics only those that are meaningful from the point of view of consciousness. Consciousness is thus concerned with establishing the patterns of meaning within which physical reality manifests itself; and quantum physics presents precisely the appropriate amount of indeterminism and flexibility to allow consciousness to do this.

The Decoherence Debate

What is decoherence?

We have seen how, on the one hand, quantum theory has many suggestive pointers to human experience, and, on the other hand, human experience offers quantum theory a way whereby it can become a complete theory. But all this depends on the possibility of the distinctive features of quantum theory (nonlocality and quantum logic) being manifested at length scales relevant to the brain. From the standpoint of mainstream physics, there are quite strong arguments that this cannot happen, which I will try to set out here. First, I need to explain a bit more about the quantum state.

I have said that the state both encapsulates the influence of past history, and specifies probabilities for the future. But the state has a richer structure to it than just probabilities. The latter are expressed by ordinary numbers, lying between 0 and 1. States, on the other hand, are described by *complex numbers,* which are made up of an ordinary number and a *phase,* an angle on the circle from 0 degrees to 360 degrees. The complex numbers that enter into the descriptions of states are called *amplitudes,* as distinct from the *probabilities* that describe classical statistics.

A way of understanding amplitudes is through the suggestive analogy of an electromagnetic wave. To specify the wave at any point one needs to know both its strength (a number) and the direction in which the electric field is pointing (an angle). For example, in a wave moving horizontally, the electric field will always be at right angles to the wave, but it might be pointing horizontally, vertically, or at any angle in between.

The distinction between probabilities and amplitudes gives rise to a distinction between a *mixture* and a *superposition.* If I toss a coin and conceal it, the probabilities of getting a head or a tail (both 0.5) can be expressed by a classical state which is a mixture of a head and a tail. There is nothing quantum mechanical about this. By contrast, consider the situation of an electron, which behaves like a little magnet whose North pole is pointing in a specific direction. I could prepare a state which is a mixture of states where the directions of the pole is up in one and down in the other, just like a tossed coin, expressing a simple lack of information. Or alternatively, I can prepare a *superposition* of the two with equal probabilities. As well as giving rise to probabilities for pole-up or pole-down, in addition it corresponds to the electron pointing in a particular horizontal direction, depending on the relative phases of the two states being combined. (This appearance of a new property is a bit like the scrambled eggs in the culinary example above.)

Can we prepare a state of a brain, or even part of a brain, which is in a superposition of different thoughts rather than a mixture? Or, can we prepare a superposition of my being in the bathroom or in the bedroom, which I mentioned earlier? Ideas about the quantum mind would require us to be able to prepare superpositions in the case of thoughts in the brain, but not in the case of the positions of material bodies. Where does the difference lie?

An answer is provided by decoherence theory, in terms of the way that the environment interacts with a superposition so as to affect the phase of the amplitudes of its constituent states. This phase is exquisitely sensitive to perturbations: the impact of a single one of the weakest photons found in nature will set the relative phase of a superposition spinning with a frequency of 400 Gigahertz (computer speeds are measured in a few Gigaherz). All that is required for this to happen is that the two states being superposed are sufficiently different for them to interact with the perturbation in different ways. This effect is called *decoherence*, from the idea that in the absence of perturbation the phases of the states remain 'coherent', in step. The bigger the system, the more likely it is to be perturbed. In the case of the direction of the North pole of an electron, a pole-up electron is virtually indistinguishable from a pole-down electron and both states respond to perturbations in the same way, so that decoherence happens very slowly At the other extreme, if we are considering a large body in two different positions, then the slightest perturbing factor will produce large effects which are quite different for the two states, leading to very rapid decoherence indeed.

As a result, even if we could prepare a large body in a superposition of states in which it had different positions, the state would almost immediately turn into a superposition whose phase had become completely random and unknown. In other words, the one factor that distinguishes a quantum superposition from a classical mixture has become inaccessible to any observation. A superposition with a random phase is the same thing as a mixture. What is the case with superposing brain states corresponding to different thoughts? These involve different distributions of various chemicals in the brain and so, one might suppose, should easily fall prey to decoherence.

Many physicists would take the view that this kills the whole idea of the quantum mind stone dead. Why, they might ask, did I not start with this section and thereby save the trouble of writing all the rest? There are two answers. First, because so much would be gained in terms of a unification of psychology and physics if a mechanism could be found for bypassing decoherence, that it is worth turning quite a few stones to try to find such a thing. Second, if we cannot get the idea of quantum mind to work, then the incompleteness of quantum theory remains a problem. This is because, though decoherence explains very nicely why, *given* that the world consists of distinct macroscopic systems, these systems then behave in a classical way (so that I cannot be superposed between the bathroom and the bedroom), it does not explain why these distinct objects should be selected from the range that is allowed by the formalism of quantum mechanics in the first place. I want therefore to outline two possible arguments which suggest that the effect of decoherence on the brain might not undermine the quantum mind.

The theory of Hameroff and Penrose (1996)

The core of this idea is to identify structures in the brain that are both very well protected from decoherence, and also suitable for taking part in the process of thinking. These are the *microtubules*, a network of long thin tubes that is part of the interior 'skeleton' of every living cell. The walls are built of regular arrays of a protein that has two stable forms, differing slightly in the way in which the units of which it is made are positioned. Superpositions are formed between states involving different arrangements of these two forms among all the proteins making up the tube. Protection from decoherence is provided by the very small

difference between the two forms of the protein, and by shield-
ing that takes place through the arrangement of water and other
molecules outside the microtubule. The microtubules are conjec-
tured to take part in thinking through the way in which waves
running up and down the tubule can coordinate computation
processes, using the two protein forms as the 'bits' in an information system. They apply this within the framework of Penrose's
theory, a state-based theory in which the state collapses as a
result of quantum gravitational processes, and where this collapse forms the 'stuff' of consciousness. The critical question
therefore becomes the calculation of whether the Penrose collapse occurs before or after the point where the state is reduced
to a mixture by decoherence. Recently Faber and Rosa (2004)
have argued that decoherence actually takes place before the
Penrose process, but that its timescale is still sufficiently long for
the idea of the quantum mind still to be viable.

The microtubules idea is promising, but there is a wealth of
work still to be done before it can become clear whether it is adequate for the issues that I have raised here. How exactly are the
computational processes in the microtubules linked into the
much larger-scale neural processes that are normally regarded
as the sole basis of thinking? Can this linkage be achieved sufficiently strongly for the microtubule processes to play a significant role in thinking, but not so strongly that decoherence
becomes too fast? If the Hameroff-Penrose theory is correct, then
the quantum computational side of the brain will take place *prior*
to collapse and hence prior to the intervention of consciousness.
This is thus contrary to my proposal, based on subjective experience, that quantum logic is in fact a characteristic feature of *conscious* mental operations, on time scales that are much longer
than those associated with microtubule processes. At the early
stage that this theory is at, however, these questions are more in
the nature of research proposals than objections.

Coherence, life and consciousness

If we adopt Penrose's theory of consciousness, then a 'theory of
mind' in the sense of the previous section is not required. Consciousness arises not from mind in the sense of the function of a
particular biological structure, but from a physical effect that is
very widely distributed in the universe. It is, however, dependent on a future theory of quantum gravity which many regard as

speculative. If this theory fails, then we can still use microtubules as a place free from decoherence, but now as part of the approach just described, using histories and a theory of mind. But how viable is this?

In previous work (Clarke, 2002) I have drawn on the idea of Mae-Wan Ho (1998), that all living organisms are 'coherent'; and I have linked this with philosophy of panpsychism, proposing that all living things are, in some possibly rudimentary way, conscious. The problem with this is that Ho uses the term 'coherent' in quite a broad sense. A confusing consequence is that 'coherent' is not the opposite of 'decoherent'! An important illustration of this comes from the example that was given publicity by Ian Marshall and Danah Zohar (1990) in which a quantum mechanical mechanism (first considered by Fröhlich) was proposed which could link together oscillating molecules over the whole brain into a single vibration that formed the substratum for consciousness. This (an instance of a general mechanism called Bose condensation) is certainly possible in principle, though there is no direct evidence that neurons actually have the mechanisms to implement this in the way that was originally proposed (see Clarke, 1994). However, suppose that a global synchronisation of this form could be achieved by some mechanism or other. It would be coherent in the sense that all the participating molecules were vibrating in phase with each other. But it need not be coherent in the sense of decoherence theory: where a Bose condensation is maintained by active physiological processes of the form discussed by Frohlich, it is still acted upon by the environment so as to turn superpositions into mixtures.[5] This could form an obstacle to any attempt to co-ordinate microtubule collapse into a phenomenon occurring across the brain.

I believe that the way ahead here is to take seriously the lessons from philosophy of the previous sections. I have argued that the inception of quantum theory has thrown into question the whole classical notion that ultimate reality lies in atomism, and I have drawn from Chalmers' arguments the idea that consciousness itself (in the sense of the word used by Chalmers and Velmans) needs to be taken as a primitive ingredient of our theory lying outside physics. This is not, however, reflected in the

---

[5] The situation is different from that of the low-temperature Bose condensations being investigated in laboratories today, where the conditions for coherence and lack of decoherence coincide.

arguments I have cited for calculating decoherence effects. These are all founded squarely on atomistic principles, taking disconnected particles as fundamental and therefore assuming that no correlations can exist unless they can be derived in a bottom-up way from particles. The role of context is ignored, and all arguments are carried out as if in the ultimate classical context of an entire universe filled with particles. The alternative, now gradually coming to the fore in philosophy, is a *participative* world view (Heron & Reason, 1997; Ferrer, 2003) where the 'objective' world is derived from the mutual overlappings of the subjective worlds of innumerable conscious beings, and coherence is inherited from consciousness, rather than consciousness being derived from coherence. Needless to say, the articulation of this within physics would constitute a difficult and radical programme for the future.

An Alternative Story of World and Self

We have now escaped from the stranglehold of Cartesian dualism which had led to the story of a meaningless universe. Descartes described the separation and interaction between mind (thinking stuff) and matter (extended stuff) in terms of their having different functions and being different substances. Such a division, as we have seen, is not viable. We have introduced the possible foundations of a unified account based on consciousness, from which objective physical entities are derived. In this the distinction between the role of consciousness in general and the specific physical structures that arise from it is made not in terms of dynamical function, but in terms of first-person meaning and third-person dynamics, respectively.

The answer to 'what is the world really like?' is a story that is as much about our self as about the world, because in this picture we play a part, in coordination with all other conscious beings and with the influence of the context of the entire universe, in shaping what the world is. We are co-creators in the universe — co-creators with God, if one chooses to use theistic terminology.

The conscious 'I' that emerges from this picture is grounded in the physical body but not identical to it. On the one hand, my consciousness is linked only with a subsystem of the many interacting and fully overlapping subsystems that make up my whole body. On the other hand, this subsystem will, through quantum non-locality, also include within it aspects of all the other beings that I am perceiving, and of people with whom I have an empathic connection.[6] We are not separated from each other and from the world, as are the atoms in the Newtonian system, but integrally connected with each other.

We have a deeply held intuition that freedom is an integral part of our being. In the light of this picture, it appears that to some extent this is illusory. When I choose between definite alternatives set out in advance, my choice is probably partly the application of comparatively mechanistic problem-solving techniques, and partly randomness. It is still my choice, in the sense that it arises from the whole of who I am, which has built up through the whole of my life, but it is not quite what I like to suppose when I talk about my 'free will'. On the other hand, when I creatively move into a new way of seeing things, a new framework of meaning, then I am changing the way my consciousness is selecting within the histories that I take part in. This lies outside the mechanistic dynamics of physics and goes to the core of the self. Ideas of value and responsibility flow from this source of creativity.

It is clear that this is a picture that affirms our humanity rather than denying it. It is also a picture that is consistent with existing science, rather than one that tries to reinvent an alternative New Age science. On the basis of this sort of approach, it is possible for us to reaffirm the values of our humanity and of our connectivity with the world around us, while at the same time building on all we have learnt through the rigorous application of science. I believe that it is only with this combination of the subjective dimension and established scientific knowledge that we will find a future on this planet.

References

Bohm, David and Hiley, Basil J. (1993), *The Undivided Universe*. London: Routledge.

Chalmers, David J. (1995) 'Facing up to the problem of consciousness', *Journal of Consciousness Studies* **2** (3), pp. 200–219.

Clarke, C.J.S. (1994), 'Distributed pumped dipole systems do not admit true Bose condensations,' *J Phys A*, **27**, pp. 5495–5500.

---

[6] See also Sheldrake (2003). The theory proposed here, however, remains within comparatively conventional physics and psychology and without the Cartesian dualism of Sheldrake's concept of mind.

Clarke, C.J.S. (2002), *Living in Connection*. Creation Spirituality Books.

Cleve, Felix M. (1965), *The Giants of Pre-Sophistic Greek Philosophy: An Attempt to Reconstruct Their Thoughts* (2 vols). Martinus Nijhoff Publishers.

Donald, Matthew J. (1995), 'A mathematical characterisation of the physical structure of observers', *Foundations of Physics*, **25**, pp. 529–571.

Faber, Jean and Rosa, Luiz Pinguelli (2004), 'Quantum models of mind: Are they compatible with environment decoherence?', quant-ph/0403051.

Ferrer, Jorge (2003), *Revisioning TranspesonalTheory*. New York: SUNY.

Hameroff, S.R. and Penrose, R. (1996), 'Orchestrated reduction of quantum coherence in brain microtubules: a model for consciousness?' In: *Toward a Science of Consciousness: The First Tucson Discussions and Debates*, ed. Hameroff, Kaszniak and Scott. MIT Press, pp. 507–540.

Heron, John and Reason, Peter (1997), 'A participatory inquiry paradigm', *Qualitative Inquiry*, **3**, pp. 274–294.

Ho, Mae-Wan (1998), *The Rainbow and the Worm: the Physics of Organisms*. World Scientific, 2nd Edn.

Kochen, Simon and Specker, Ernst (1967), 'The problem of hidden variables in quantum mechanics', *Journal of Mathematics and Mechanics*, **17**, pp. 59–87.

Murphy, Nancey and Ellis, George (1996), *On the Moral Nature of the Universe: Theology, Cosmology and Ethics*. Fortress Press.

Nagel, Thomas (1979), 'What is it like to be a bat?' in Thomas Nagel, *Mortal Questions*. Cambridge University Press.

Sheldrake, Rupert (2003), *The Sense of Being Stared at and Other Aspects of the Extended Mind*. Hutchinson.

Zohar, Danah (1990), *The Quantum Self*. London: Bloomsbury.

Ravi Ravindra

Yoga, Physics and Consciousness

Introduction

In spite of our wish to reconcile science and spiritual insight, we are very far from even having clear questions to raise about these two approaches to reality. We wish these disciplines to be reconciled because they both appear to us to be significant and profound manifestations of the human psyche, and we imagine that somehow in modern times we have found a reconciliation. Both yoga, which is an expression of spiritual insight, and physics are interested in objective knowledge. However, the two 'knowledges' are different from each other. We need to be aware of these differences if we are to avoid settling for an easy integration or a superficial reconciliation. Nothing is more misleading than to imagine that there is peace when there is no peace. The illusion that we have already found what we need will prevent us from seeking further.

Science assumes an abstract and purely rational construct underlying perceived reality. So what is experienced is called 'appearance', and the mental construct is labeled 'reality'. The scientific pursuit speculates about the imagined reality and puts these speculations to experimental tests, which involve only certain limited perceptions. The so-called objective reality of scientific concern is in fact a conjecture — perhaps one of many that are possible. However — and this is where the importance and glory of science lie — these subjective projections of the mind are confirmed or falsified by inter-subjective experimental procedures.

Nevertheless, testing procedures are not wholly independent of the theoretical framework in which the observations are made. As scientific experiments become more and more elaborate, whether an observation is taken to be a confirmation of a given conjecture is increasingly a matter of interpretation. It is not possible to make a scientific observation without a prior theoretical system. In science, any theory is better than no theory. Theorising is fundamental to scientific activity; what scientists subject to experimental observations is not nature, but their conjectures about nature.

In an argument with Albert Einstein, Niels Bohr said, 'It is wrong to think that the task of physics is to find out how nature is. Physics concerns what we can say about nature.' (Moore, 1966, p. 406.) The scientific revolution marks a shift not only from experience to experiment,[1] but also from seeking certain truth to theorising about probable truths. In science, reality is theory.

Reality discovered through science is not necessarily something that is given, which we try to perceive more and more clearly and comprehensively by deepening or cleansing our perceptions, as one attempts, for example, in yoga. It is instead something postulated on the basis of data gathered through our ordinary perceptions, or perceptions that have been quantitatively extended through scientific instruments, but not qualitatively transformed.

The scientific assumption about human beings is that they are essentially rational cognisers, and that everything else about them is secondary and capable of explanation in terms of their basic rational nature. This view of a person as primarily a passionless, disembodied mind, which would be recognised as the rigorously intellectual point of view, is shared by all who claim to be scientific in their professional work, from Descartes to the modern analytical philosophers. Other human faculties — feelings and sensations — are not considered capable of either producing or receiving real knowledge. It is no doubt true that, as we are, our ordinary sensory and emotional experiences are limited and subjective. In science, an attempt is made to minimise

---

[1]   See Ravindra (2002).

the dependence on such perceptions by agreeing that the corresponding aspects of reality not be considered as objectively real and by dealing with only those aspects to which rational constructs can be applied.

The task of yoga, and of all spiritual disciplines, is not the same as the task of the scientific inquiry. Whereas science seeks to understand and control processes in the world, using the rational mind as the tool of exploration and explanation, yoga seeks to transform the human being so that the reality behind the world can be experienced.

According to Patañjali, the author of the classical text on yoga, 'Yoga is the quieting of the *vrittis* (projections, turnings, movements, fluctuations) of the mind. Then the true or essential form of the seer is established. Otherwise, there is identification with the projections.' (*Yoga Sutra*, 1.2–4.) *Vrittis* of the mind, like Plato's shadows in the cave, are chimeras, taken to be real. For Patañjali, the mind needs to be completely quiet in order to know the truth about anything. The quiet mind is the original state. However, there are obstacles (*kleshas*) which prevent one from seeing the truth. The *Yoga Sutra* speaks about what these *kleshas* are and about how to remove them. Patañjali's yoga is a teaching to reach the still mind — one's true nature. Only then can true knowledge about anything be obtained.

It should be stressed right at the outset that the point of view informing the theory and practice of yoga originates from above, that is to say, from the vision of the highest possible state of consciousness. It is not something which has been forged or devised from below, or which can even be understood by the human mind, however intelligent such a mind may be. Yoga is a supra-human (*apaurusheya*) revelation; it is from the realm of the gods. Mythologically, it is said that the great God Shiva taught yoga to his beloved Parvati for the sake of humanity. It cannot be validated or refuted by human reasoning; on the contrary, the relative sanity or health of a mind is measured by the extent to which it accords with what the accomplished sages, who have been transformed by the practice of yoga, say. It is a vision from the *third eye*, relative to whose reality the two usual eyes see only shadows.

However, it is important to emphasise that no mere faith, and certainly nothing opposed to knowledge, is needed in yoga. What is in fact required is the utmost exertion of the whole of the human being — mind, heart, and body — for the practice which would lead to a total transformation of being, a change not less than in a species mutation. Yoga brings the vision from the third eye of Shiva and of the sages for us to receive, and aims at helping us develop and open the third eye in ourselves so that we may see with the spiritual vision of Shiva and of the sages. The etymology of the word *yoga* — derived from the root *yuj*, meaning 'to yoke, unite, harness' — conveys the aim of yoga which is union with the highest level. When the human body-mind is harnessed to the Spirit (*Purusha, Atman, Brahman*), which is as much within a human being as outside, the person is in yoga. In that state, the person is free of all *kleshas* and sees the way it is.

The fulfilment of the aim of yoga requires the transformation of a human being from the natural and actual form to a perfect and real form. The *prakrita* (literally, natural, vulgar, unrefined) state is one in which a person compulsively acts in reaction to the forces of *prakriti* (nature, causality, materiality) which are active both outside as well inside a person. Ordinarily, a person is a slave of the mechanical forces of nature and all actions are determined by the *Law of Karma*, the law of action and reaction. Through yoga one can become *samskrita* (literally well-formed, cultured, refined), and thus no longer be wholly at the mercy of natural forces and inclinations. The procedure of yoga corresponds to the process of education. It helps to draw out what is, in fact, already present, but which is not available. The progressive bringing out of The Real Person (*Purusha*) in an aspirant is much like the releasing of a figure from an unshaped stone. A remark of Michelangelo 'I saw an angel in the block of marble and I just chiseled and chiseled until I set him free'.

The undertaking of yoga concerns the entire person, resulting in a reshaping of the mind, the body, and the emotions; in short in a *new birth*. The yogi — the one who practices yoga and who is transformed through this practice — is the artist, the stone, and the tools. But unlike in sculpting, the remoulding involved in yoga is essentially from the inside out. Lest this analogy be misunderstood to suggest that yoga leads to a rugged individualism in which individuals are the makers of their own destiny, it should be remarked that the freedom that a yogi aspires to is less a freedom *for* the self, and more a freedom *from* the self. From a strict metaphysical point of view, yogis cannot be said to be the artist of their own lives; the real initiative belongs only to *Brah-*

man who is lodged in the heart of everyone. A person does not create a state of freedom; but with a proper preparation, an individual can let go of an attachment to the surface and to the insistence to possess and control everything, in order to be possessed by what is deep within.

Yoga aims at *moksha* which is unconditioned and uncaused freedom. This state of freedom is, by its very nature, beyond the dualities of being–nonbeing, knowledge–ignorance, and activity–passivity. The way to *moksha* is yoga, which serves as a path or a discipline towards integration. Yoga is as much *religion*, as *science*, as well as *art* since it is concerned with being (*sat*), knowing (*jñana*) and doing (*karma*). The aim of yoga, however, is beyond these three, as well as beyond any opposites that they imply.[2]

The Body and the Embodied

Yoga begins from a recognition of the human situation. Human beings are bound by the laws of process and they suffer as a consequence of this bondage. Yoga proceeds by a focus on knowledge of the self. Self-knowledge may be said to be both the essential method and the essential goal of yoga. However, self-knowledge is a relative matter. It depends not only on the depth and clarity of insight, but also on what is seen as the *self* to be known. A progressive change from the identification of the self as the body (including the heart and the mind) to the identification of the self as inhabiting the body is the most crucial development in yoga. Ancient and modern Indian languages reflect this perspective in the expressions used to describe a person's death: in contrast to the usual English expression of *giving up the ghost*, one *gives up the body*. It is not the body that has the Spirit, but the Spirit that has the body. The yogi identifies the person less with the *body* and more with the *embodied*.

The identification of the person with something other than the body-mind and the attendant freedom from the body-mind is possible only through a proper functioning and restructuring of the body and the mind. Here it is useful to retain the Sanskrit word *sharira* in order to steer clear of the modern Western philosophic dilemma called the 'mind–body' problem. Although *sharira* is usually translated as body, it means the whole psycho-

---

[2]   In this connection, see Ravindra (1978).

somatic complex of the body, mind, and heart.[3] *Sharira* is both
the instrument of transformation as well as the mirror indicating
it. Knowing the way a person sits, walks, feels, and thinks, can
help in knowing the relatively 'realer' self; the knowing of this
self is then reflected in the way a person sits, walks, feels, and
thinks. *Sharira*, which is individualised *prakriti,* is the medium
necessary for the completion and manifestation of the inner spiritual being, which itself can be understood as individualised
Brahman (literally, the Vastness) whose *body* is the whole of the
cosmos, subtle as well as gross. There is a correspondence
between the microcosmos which is a human being, and the
macrocosmos. The more developed a person is, the more the person corresponds to the deeper and more subtle aspects of the cosmos — only a fully developed human being (*Mahapurusha*)
mirrors the entire creation. To view the *sharira* or the world as a
hindrance rather than an opportunity is akin to regarding the
rough stone as an obstruction to the finished figure. *Sharira* is the
substance from which each one of us makes a work of art, according to our ability to respond to the inner urge and initiative. This
substance belongs to *prakriti* and includes what are ordinarily
called psychic, organic, and inorganic processes. The view that
mind and *body* follow the same laws, or the fact that the *psychic,
organic,* and *inorganic* substances are treated alike, does not lead
to the sort of reductionism associated with the modern scientific
mentality in which the ideal is to describe all of nature ultimately
in terms of dead matter in motion reacting to purposeless forces.
Prakriti, although following strict causality, is alive and purpose-
ful, and every existence, even a stone, has a psyche and purpose.

Seeing Through the Organs of Perception

Although there are many kinds of yogas, such as *karma yoga*
(integration through action), *bhakti yoga* (union through love),
jñana yoga (yoking through knowledge), and others, the Indian
tradition has in general maintained that there is only one central

[3]   *Sharira* here has the same import as *flesh* in the *Gospel According to St. John,*
for example in *John* 1:14 where it is said that 'The Word became flesh and
dwelt in us.' In this connection, see Ravindra (1990). The important point,
both in the Indian context and in *John* is that the spiritual element, called
Purusha, Atman, or *Logos (Word)* is above the whole of the psychosomatic
complex of a human being, and is not to be identified with *mind.*

yoga, with one central aim of harnessing the entire body-mind to the purposes of the Spirit. Different yogas arise owing to varying emphases on methods and procedures adopted by different teachers and schools. The most authoritative text of yoga is regarded to be the *Yoga Sutra,* which consists of aphorisms of yoga compiled by Patañjali sometime between the second century BCE and the fourth century CE from material already familiar to the *gurus* (teachers) of Indian spirituality. It is clearly stated by Patañjali that clear seeing and knowing are functions of *purusha* (the inner person) and not of the mind. The mind is confined to the modes of judgment, comparison, discursive knowledge, association, imagination, dreaming, and memory through which it clings to the past and future dimensions of time. The mind with these functions and qualities is limited in scope and cannot know the objective truth about anything. The mind is not the true knower: it can calculate, make predictions in time, infer implications, quote authority, make hypotheses or speculate about the nature of reality, but it cannot see objects directly, from the inside, as they really are in themselves.

In order to allow direct seeing to take place, the mind, which by its very nature attempts to mediate between the object and the subject, has to be quieted. When the mind is totally silent and totally alert, both the real subject (*purusha*) and the real object (*prakriti*) are simultaneously present to it. When the seer is there and what is to be seen is there, seeing takes place without distortion. Then there is no comparing or judging, no misunderstanding, no fantasising about things displaced in space and time, no dozing off in heedlessness nor any clinging to past knowledge or experience; in short, there are no distortions introduced by the organs of perception, namely the mind, the feelings, and the senses. There is simply the *seeing* in the present, the living moment in the eternal now. That is the state of perfect and free attention, *kaivalya,* which is the aloneness of seeing, and not of the seer separated from the seen, as it is often misunderstood by commentators on yoga. In this state, the seer sees through the organs of perception rather than with them.

It is of utmost importance from the point of view of yoga to distinguish clearly between the mind (*chitta*) and the real Seer (*purusha*). *Chitta* pretends to know, but it is of the nature of the known and the seen, that is, it is an object rather than a subject. However, it can be an instrument of knowledge. This misidentification of the seer and the seen, of the person with his organs of perception, is the fundamental error from which all other problems and sufferings arise (*Yoga Sutra* 2:3-17). It is from this fundamental ignorance that *asmita* (I-am-this-ness, egoism) arises, creating a limitation by particularisation. *Purusha* says 'I AM'; *asmita* says 'I am this' or 'I am that.' From this egoism and self-importance comes the strong desire to perpetuate the specialisation of oneself and the resulting separation from all else. The sort of knowledge which is based on this basic misidentification is always coloured with pride, a tendency to control or fear.

The means for freedom from the fundamental ignorance which is the cause of all sorrow is an unceasing vision of discernment (*viveka khyati*); such vision alone can permit transcendental insight (*prajña*) to arise. Nothing can force the appearance of this insight; all one can do is to prepare the ground for it; it is the very purpose of *prakriti* to lead to such insight, as that of a seed is to produce fruit; what an aspirant needs to do in preparing the garden is to remove the weeds which choke the full development of the plant. The ground to be prepared is the entire psychosomatic organism, for it is through that organism that *purusha* sees and *prajña* arises, not the mind alone, nor the emotions nor the physical body by itself. One with dulled senses has as little possibility of coming to *prajña* as the one with a stupid mind or hardened feelings. Agitation in any part of the entire organism causes fluctuations in attention and muddies the seeing. This is the reason why in yoga there is so much emphasis on the preparation of the body for coming to true knowledge. It is by a reversal of the usual tendencies of the organism that its agitations can be quieted, and the mind can know its right and proper place with respect to *purusha*: that of the *known* rather than the *knower*. (*Yoga Sutra* 2:10; 4:18-20.)[4]

Samyama — Attention as the Instrument of Knowledge

In classical Yoga, there are eight limbs: the first five are concerned with the purification and preparation of the body, emotions, and breathing and with acquiring the right attitude; the last three limbs are called inner limbs compared with the first

---

[4]   In this connection, see. Ravindra (1989).

five which are relatively outer. The last three are *dharana, dhyana,* and *samadhi. Dharana* is concentration in which the consciousness is bound to a single spot. *Dhyana* (from this word is derived the Japanese *Zen* through the Chinese *Ch'an* and Korean *Sôn*) is the contemplation or meditative absorption in which there is an uninterrupted flow of attention from the observer to the observed. In these the observer acts as the centre of consciousness which sees. When that centre is removed, that is to say when the observing is done by *purusha*, through the mind emptied of itself, that state is called *samadhi* — a state of silence, settled intelligence, and emptied mind, in which the mind becomes the object to which it attends, and reflects it truly, as it is.

The insight obtained in the state of *samadhi* is truth-bearing (*ritambhara*); the scope and nature of this knowledge is different from the knowledge gained by the mind or the senses. The insight of *prajña* reveals the unique particularity, rather than an abstract generality, of an object. Unlike a mental knowledge, in which there is an opposition between the object and the subjective mind, an opposition that inevitably leads to sorrow, the insight of *prajña*, born of the sustained vision of discernment, is said to be the deliverer. This insight can be about any object, large or small, far or near; and any time, past, present, or future, for it is without time-sequence, present everywhere at once, like a photon in physics in its own frame of reference.

The Natural Science of Yoga

It is wrong to suggest that yoga is not interested in the knowledge of nature and is occupied only with self-knowledge. From the perspective of yoga, this is an erroneous distinction to start with, simply because any self, however subtle, that can be known is a part of nature and is not distinct from it. The deepest self, to which alone belongs true seeing and knowing, cannot be known; but it can be identified with. One can become that self (*Atman, Purusha*) and know with it, from its level, with its clarity. In no way is *prakriti* considered unreal or merely a mental projection; she is very real, and though she can overwhelm the mind with her dynamism and charms and veil the truth from it, yet in her proper place and function she exists in order to serve the real person *(purusha)*.

However, it is certainly true that the procedures, methods, attitudes, and perceptions involved in yoga are radically different from those in modern science, as are the aims of the two types of knowledge.[5] In a summary way, one can say that in contradistinction to modern science the knowledge in yoga is a third eye knowledge, transformational in character. It is a knowledge which does not bring violence to the object of its investigation; it is a knowledge by participation, rather than by standing apart or against the object. Knowledge in yoga is primarily for the sake of true seeing and the corresponding freedom.

The basic research method of the science of nature according to yoga is to bring a completely quiet mind and to wait without agitation or projection, letting the object reveal itself in its own true nature, by colouring the transparent mind with its own colour. This science is further extended by the principle of analogy and isomorphism between the macrocosmos and the microcosmos which is the human organism. Therefore, self-knowledge is understood to lead to a knowledge of the cosmos. An example of this isomorphism is to be found in the *Yoga Darshana Upanishad* (4:48-53) where the external *tirtha* (sacred ford, places of pilgrimage, holy water) are identified with the various parts of the organism: 'The Mount Meru is in the head and Kedara in your brow; between your eyebrows, near your nose, know dear disciple, that Varanasi stands; in your heart is the confluence of the Ganga and the Yamuna.'

A large number of aphorisms in the *Yoga Sutra* (3:16-53) describe the knowledge and the powers gained by attending to various objects in the state of *samyama*. The three inner limbs of yoga, namely, *dharana*, *dhyana*, and *samadhi*, together constitute what is called *samyama* (discipline, constraint, gathering). It is the application of *samyama* to any object which leads to the direct perception of it, because in that state the mind is like a transparent jewel which takes on the true colour of the object with which it fuses (*Yoga Sutra* 1:41). The special attention which prevails in the state of *samyama* can be brought to bear on any thing which can be an object of perception, however subtle, that is, on any aspect of *prakriti*. For example, we are told that, through *samyama* on the sun, one gains insight into the solar system, and, by *samyama* on the moon, knowledge of the arrangement of the

[5]   In this connection, see Ravindra (1980).

stars (*Yoga Sutra* 3:26-27). Similarly, many occult and extraordinary powers (*siddhis*) accrue to the yogi by bringing the state of
samyama to bear upon various aspects of oneself: for example, by
samyama on the relation between the ear and space, one acquires
the divine ear by which one can hear at a distance or hear
extremely subtle and usually inaudible sounds. Many other
powers are mentioned by Patañjali; however, none of them are
his main concern. There is no suggestion that there is anything
wrong with these powers; no more is there a suggestion that
there is something wrong with the mind. The point is more that
the mind, as it is, is an inadequate instrument for gaining true
knowledge; similarly, these powers, however vast and fascinating, are inadequate as the goal of true knowledge.

Necessity of Transformation

It cannot be said too often that higher levels cannot be investigated by, or from, a lower level. What can be studied by the mind
in the modern scientific mode is only that which can in some
senses be manipulated and controlled by the mind and is thus
below the level of the mind. In the presence of something higher
than itself, the mind needs to learn how to be quiet and to listen.

Another remark needs to be made about the various practices
of yoga: what is below cannot coerce what is above. One cannot
force higher consciousness or Spirit by any manipulation of the
body, mind or breath. A right physical posture or moral conduct
may aid internal development but it does not determine it or
guarantee it. More often external behaviour reflects internal
development. For example, a person does not necessarily
become wise by breathing or thinking in a particular way; but a
person breathes and thinks in that way because he or she is wise.
Actions reflect being more than they affect it.

A very important heuristic principle in modern science inter-
feres with the knowledge of a radically different and higher
level. This principle enters as the Copernican Principle in
Astronomy and Cosmology and as the Principle of
Uniformitarianism in Geology and Biology, one to do with space
and the other with time. According to the former, any point in
the universe can be taken to be the centre, for in each direction
the universe on the large is homogeneous and isotropic. The latter principle says essentially that the same laws and forces have

operated in the past as in the present. Neither of these principles have anything to say about levels of consciousness. But in practice one consequence of these principles has been a denial of a radical difference not only in terms of regions of space and time, but also in terms of levels of being among humans. One of the important aspects of modern science, starting with the great scientific revolution of the sixteenth and seventeenth centuries, has been a scientifically very successful idea that the materials and laws on other planets and galaxies, and in the past and future times, can be studied in terms of the laws, materials and forces available to us now on the earth. But, almost by implication and quite subtly, this notion has done away with the analogical and symbolic modes of thinking according to which a fully developed person could mirror internally the various levels of the external cosmos.

A Science of Consciousness Requires Transformed Scientists

When the ancients and even the medieval thinkers in Europe, China or India — in their sciences of alchemy, astronomy and cosmology — spoke of different planets having different materials and different laws, at least in part it meant that various levels of being or consciousness have different laws. From this perspective higher consciousness cannot be understood in terms of, or by, a lower consciousness. The subtler and higher aspects of the cosmos can be understood only by the subtler and higher levels within humans. True knowledge is obtained by participation and fusion of the knower with the object of study, and the scientist is required to become higher in order to understand higher things. As St. Paul said, things of the mind can be understood by the mind; things of the spirit by the spirit. The ancient Indian texts say that only by becoming Brahman can one know Brahman. The *Gandharva Tantra* says that 'no one who is not himself divine can successfully worship divinity.' For Parmenides and for Plotinus 'to be and to know are one and the same'.[6]

This has implications for any future science of higher consciousness which would hope to relate with what is real. Such a science would have to be *esoteric*, not in the sense of being an

----

[6]   Parmenides, *Diels, Fr.* 185; Plotinus, *Enneads* vi. 9.

exclusive possession of some privileged group, but because it would speak of qualities which are more subtle and less obvious, such a science would demand and assist the preparation, integration and attunement of the body, mind and heart of the scientists so that they would be able to participate in the vision revealed by higher consciousness. In the felicitous phrase of Meister Eckhart, one needs to be 'fused and not confused.' *Tatra prajña ritambhara* (there insight is naturally truth-bearing), says Patañjali in the *Yoga Sutra* (1.48-49; 2.15; 3.54). This preparation is needed in order to open the third eye, for the two usual eyes do not correspond to the higher vision. It is only the third eye that can see the hidden Sun, for as Plotinus says, 'to any vision must be brought an eye adapted to what is to be seen, and having some likeness to it. Never did the eye see the sun unless it had first become sun-like, and never can the soul have vision of the First Beauty unless itself be beautiful.' (*Enneads* I. 6.9.)

The important lesson here from the perspective of any future science of consciousness is the importance of knowledge by identity. We cannot remain separate and detached if we wish to understand. We need to participate in and be one with what we wish to understand. Thus Meister Eckhart:

> Why does my eye recognise the sky, and why do not my feet recognise it? Because my eye is more akin to heaven than my feet. Therefore my soul must be divine if it is to recognise God.[7]

Similarly Goethe:

> Waer' nicht das Auge sonnenhaft,
> Die Sonne könnt' es nie erblicken.
> Laeg' nicht in uns des Gottes eigene Kraft,
> Wie koennt' uns Göttliches entzücken?

> If the eye were not sensitive to the sun,
> It could not perceive the sun.
> If God's own power did not lie within us,
> How could the divine enchant us?

In the well nigh universal traditional idea of a correspondence between a human being and the cosmos, the microcosmos-macrocosmos homology, it is easily forgotten that this idea does

---

[7]   Quoted by Klaus K. Klostermaier (1994), footnote no. 20, p. 533.

not apply to every human being. It is only the fully developed person (*Mahapurusha*) who is said to mirror the whole cosmos. Such developed persons are quite rare. The idea of inner levels of being (or of consciousness) is absolutely central, as is the question 'What is a person?' It is difficult to convince oneself that the various spiritual disciplines for the purpose of transformation of human consciousness can be dispensed with by developing concepts or instruments from relatively lower levels of consciousness. But unwillingness to accept the need for radical transformation and to subject oneself to a spiritual discipline is ubiquitous. Even when the idea of transformation has an appeal, one wishes to be transformed without changing — without a renunciation of what one now is and with an attitude of saying, 'Lord, save me while I stay as I am.'

It is important to remark that it is not possible to come to a higher state of consciousness without coming to a higher state of conscience. The general scholarly bias tends to be towards a study of various levels of consciousness-which are much more often spoken of in the Indic traditions — and not so much towards various levels of conscience which are more frequently elaborated in the Biblical traditions. It would be difficult to make much sense of Dante's *Divine Comedy* without an appreciation of levels of conscience. In many languages, such as Spanish, French and Sanskrit, the word for both conscience and consciousness is the same. This fact alone should alert us to the possibility of an intimate connection between the two. The awakening of conscience is the feeling preparation for an enhancement of consciousness. It is not possible to come to a higher state of consciousness without coming to a higher state of conscience. On the other hand, those who are in touch with higher levels of consciousness naturally manifest largeness of heart. Inclusiveness and compassion bespeak a sage as a particular kind of fragrance indicates the presence of a rose.

The search for Truth — when it becomes more and more mental and divorced from deeper and higher feelings such as compassion, a sense of the oneness of the all, and the like — leads to feelings of isolation and accompanying anxiety. In this sense of isolation of oneself from all else — from other human beings as well as from the rest of nature — fear and self-importance enter. The silence of the vast spaces frightens us if we do not feel deeply that we belong to the entire cosmos. Then one wants to control others and conquer nature. Much of our modern predicament arises from this very dedication to truth in an exclusively mental manner. Feelings of alienation of our selves as isolated egos naturally follow.

The First Person Universal

In our attempts to find objective knowledge, which is the great aspiration of science, we cannot eliminate the person. What is needed in fact is an enlargement of the person — freed from the merely personal and subjective — to be inclusive. In order to comprehend one needs to be comprehensive — not as a horizontal extension of more and more knowledge, but as a vertical transformation in order to participate in the universal mind.

The well-known physicist John Wheeler summarised a profound perspective in one of his classical quips: 'It from bit.' That is to say that reality as known by us is derived from bits of information. Thus a consideration of consciousness, and various levels of it becomes immediately relevant right at the foundation of any theory of knowledge as well as Physics. Not surprisingly, this is very reminiscent of the remark of Bohr quoted earlier in this essay.

It is true that we humans know and think, the question is what or who thinks. During a conversation with the author, J. Krishnamurti said quite simply, 'You know, sir, it occurs to me that K does not *think* at all. That's strange. He just *looks*.' (See Ravindra, 2003.) We know from association that *K* was a short form of *Krishnamurti*. But what is *Krishnamurti* a short form of? Of the entire cosmos? Not him alone, potentially so each one of us. If this is true, what looks and knows through thought rather than with thought?

The purpose of spiritual disciplines such as yoga is the realisation of the First Person Universal, rather than the first person singular: the One manifests itself in myriad and quite unique forms. Only such a person can know without opposition and separation, freed from any desire to control or to manipulate. Then one loves what one knows.

References

Klostermaier, Klaus K. (1994), *A Survey of Hinduism*. State University of New York Press, second edition.

Moore, Ruth (1966), *Niels Bohr*. New York: Knopf.

Ravindra, R. (1978), 'Is religion psychotherapy? An Indian view', *Religious Studies* **14**,,pp. 389–397; reprinted in R. Ravindra: *Yoga and the Teaching of Krishna*, Theosophical Publishing House, Adyar, Chennai, India, and Wheaton, IL, 1998.

Ravindra, R. (1980), 'Perception in physics and yoga,' *Re-Vision: Jour. Knowledge and Consciousness*, **3**, pp. 36–42. Reprinted in *Science and the Sacred: Eternal Wisdom in a Changing World*. Wheaton, IL: Quest Books.

Ravindra, R. (1989), 'Yoga: The royal path to freedom,' in *Hindu Spirituality: Vedas Through Vedanta*, ed. K. Sivaraman, Vol. 6 of *World Spirituality: An Encyclopedic History of the Religious Quest* (New York: Crossroad) pp. 177–191. Reprinted in *Yoga and the Teaching of Krishna*. Wheaton, IL: Quest Books.

Ravindra, R. (1990), *The Yoga of the Christ* (Shaftesbury: Element Books) Reissued under the title *Christ the Yogi* by Inner Traditions International, Rochester, VT, 1998.

Ravindra, R. (2002), 'Experience and Experiment: A Critique of Modern Scientific Knowing,' in R. Ravindra: *Science and the Sacred: Eternal Wisdom in a Changing World*. Wheaton, IL: Quest Books.

Ravindra, R. (2003), *Centred Self without Being Self-centred: Remembering Krishnamurti*. Sandpoint, ID: Morning Light Press.

Peter Fenwick

Neurophysiology, Consciousness & Ultimate Reality

The Nature of Western Science

Western science is based on the rationalism of Descartes, Galileo, Locke, Bacon and Newton. Galileo defined a two stuff universe: matter and energy. These stuffs, he said, had primary and secondary qualities. Primary qualities were those aspects of nature that could be measured, such as velocity, acceleration, weight, mass etc: secondary qualities were the qualities of subjective experience, such as smell, vision, truth, beauty, love etc. Galileo maintained that the domain of science was the domain of primary qualities. Secondary qualities were non-scientific. 'To excite in us tastes, odours and sounds I believe that nothing is required in external bodies except shapes, numbers, and slow or rapid movements. I think that if ears, tongues and noses were removed, shapes and numbers and motions would remain but not odours or tastes or sounds.'

Western primary quality science has been outstandingly successful in examining and quantifying the world around us, and in producing our current technology, but it still, as Sherrington said, 'puts its fingers to its lips and is silent' when the question of consciousness arises. The reason for this is that consciousness, the view of the observer, is a secondary quality and thus not within the domain of science. This leaves our science very lop-sided, as only the physical aspects of any phenomenon — a 'view from nowhere' as it has been described — can be investigated by the scientific method. Yet a moment's thought, as Max Velmans has pointed out in his book *Understanding Consciousness*, shows that all described phenomena are essentially psychological entities. It is the way that the evidence is obtained that makes the difference between 'objective' and 'subjective' qualities. Objective qualities are tested by asking many different individuals if their psychological concepts, or subjective state of 'what is out there' match, e.g. do we all see the same pointer readings when we do the same experiment? Subjective qualities are tested by asking a group of people under the same conditions what their internal experiences are like. In the west a science of secondary qualities is slowly being evolved. This form of science incorporates an Eastern perspective and is used to investigate mind directly with the students examining the way their mental states change with different attentional practices. An excellent example is the *zen sesshin* which has been finely honed over the years to produce in its practitioners an expansion of consciousness.

Two major philosophical schools currently attempt to explain the nature of consciousness and tackle brain function. Dennett's (1991) neurophilosophy argues that consciousness and subjective experience are just the functions of neural nets, the brain identity theory, and there is nothing beyond this — all is only brain. At the other extreme is the philosophy of Nagel (1974) who argues that it is never possible to learn from an objective third-person point of view what it is like to have a first-person experience. Nagel argues that however much we understand about the neurophysiology of the functioning of a bat's brain, we will never know what is it like *to be* a bat. Searle (1992) argues from an intermediate position. He regards subjective experience as being a property of neural nets, but he does not agree with Dennett that a full understanding of neural net functioning is sufficient to explain subjective experience. He believes we need a Newton of neurophysiology to produce an entirely new principle. So for the present consciousness still remains unexplained, even with the expansion of neuro-imaging which has led to many correlations, there are still no explanations.

Downward Causation

Another major difficulty relates to the question of control within the central nervous system. As an experiencing human being, I feel that within limits I can control my movements, attend to a specific sensory input, and to some extent control my thinking. However, the main thrust of reductionist science is that upward causation (neuronal functioning) is the prime cause of experiential control. If that was so we would have to take a mechanistic view of human kind and there would be no question of free will, or of true creativity which is not simply mechanistic.

Science has been concerned with upward causation for so long that it is only now that it is accepting that macroscopic events within a biological system (in this case mind and meaning) may play a major part in the organisation of, and may direct the physics, chemistry and biology of, lower order systems.

This control by higher order systems of lower order systems within the body is called downward causation. Control is thought to go from mind (including social and cultural meaning) through the central nervous system to bodily function. Roger Sperry pointed out that downward causation within the central nervous system is a common property 'things are controlled not only from below upwards but also from above downwards by mental ... and other macro properties (furthermore) primacy is given to the highest level control rather than the lowest.' (Sperry 1987.)

This view, giving prominence to downward causation, helps to redress the balance and allows the driver of the brain, in certain spheres, to be the conscious individual. But again, without a theory of consciousness which links brain directly to the possibilities of conscious experience, we remain enmeshed in a reductionist trap.

The Role of Anomalies

The current scientific view is that psychological processes are generated entirely within the brain and limited to the brain and the organism. Over the last 50 years large numbers of parapsychological experiments have been carried out which suggest that mind is not limited to the brain and that it is possible to demonstrate directly the effect of mind on other minds (telepathy) and the effect of mind on matter (psychokinesis). For those interested in a more comprehensive review of this subject, the recent book by Dean Radin *The Conscious Universe* provides a wide range of references to the studies and examines some of the meta-analyses which have demonstrated these effects.

The Current Position

As a practising neuropsychiatrist, I believe that the situation now is in some respects very much better than it was when my interest in consciousness was first aroused in the 1950s. Current functional neuroimaging techniques have led to a much better understanding of the brain in action and have revealed a new phrenology of the mind. The brain appears to work as a set of interlocking modules, each one with a defined location on the cortex, and each with a specific function, all joined together in a 'magical' way (the binding problem) to produce the unified world view of conscious experience. Even with this expansion in understanding there is still no clear hypothesis pertaining to the nature of subjective experience or consciousness.

However, any way forward must involve analysis of the meta-physical foundations of science. As Professor Willis Harman argued, from the point of view of western science,

> man is torn from nature, mind from matter and science from religion. Scientists are too quick to assume that the philo-sophical premises of science are not at issue when the fact of the matter is that many debates about the nature of ultimate reality centre round fundamental ontological and epistemological questions.

The current assumptions of scientific materialism, which are usually covert and not discussed when theories of the nature of ultimate reality are being formulated, lead to a reality which has to be based only on the primary qualities of Galileo, with all the limitations that this imposes. These scientific materialistic assumptions are:

- Objectivism — knowledge is 'out there'
- Monism — there is only one 'stuff' (and it is matter)
- Universalism — it is the same everywhere
- Reductionism — provides the only reliable route to knowledge
- Causal closure principle (physical world causally closed)

- Physicalism — only physical objects exist
- Moral neutrality

This set of criteria, once they are clearly seen, can only lead to a distorted view of the world. Firstly, consciousness is excluded and it can only be added in as an extra which somehow arises. But as the whole basis of the perception of life and our understanding of the world is within consciousness, it is highly unlikely that there will be a formula which will allow it to arise unless consciousness in some way is given a part in our theories. Current neuroscience, which examines brain function in detail, comes face to face with the absence of consciousness in the brain, or indeed in any of the theories that relate to brain imaging. The best that can be established are correlations, *but correlation is not causation.*

Scientific materialism is appropriate only in areas that do not involve subjectivity. What is needed now is a new science of consciousness that does justice to subjective experience. In my view, this new science of consciousness must include a detailed role for brain mechanisms, an explanation for the action of mind outside the brain, and an explanation of free will, meaning and purpose. It should also give an explanation of wide mental states, including mystical experience and near-death experiences, when the experiencer sees through into the structure of the universe. Finally, it should provide a clear explanation for apparent downward causation (purpose) throughout the universe and in the brain, as well as some solution to the questions raised, particularly in Eastern cultures, of the survival of aspects of consciousness after death.

As Chris Clarke observes in his paper, the study of quantum mechanical effects suggests that the universe is highly interconnected and that particles interact with each other at a distance. Thus the idea that mind could also be interactive outside the skull is theoretically possible. The quantum mechanical theories of Chris Clarke and Michael Lockwood, and the quantum gravitational theories of Roger Penrose and Stuart Hameroff are all possibilities. However, to my mind, the current most likely contender to link consciousness with brain function, as it has wider explanatory power and leads to several testable predictions is a theory by Amit Goswami (1993). He argues that consciousness is a basic stuff of the universe and exists like energy. When a choice is involved, an observation made, the wave function is collapsed in consciousness and matter arises, the standard wave/particle duality theory. His major contribution to the debate is that there is only one observer, and this is a universal, undivided consciousness. He argues that brains have evolved a special mechanism for 'trapping' consciousness, so that when consciousness interacts with brain processes the probability wave collapses, on the one hand producing the external object, and on the other, subjective experience of that object.

This theory has significant explanatory power, as it will link together the binding problem within the brain, parapsychological phenomena, and more particularly for the neuropsychiatrist, a possible explanation of the wide mental states when the individual comes to experience universal love, light and consciousness, the transcendent experience. Those who experience these states say that every particle of the world is pulsing with this energy and they recognise that this is the timeless background from which all has arisen and to which all will return, perfect in its perfection and existing for ever, as time has no meaning. Of further interest, the theory suggests a mechanism for creativity — the consciousness field — which contains all possibilities and is endlessly creative and can be tapped directly by brain processes. This theory does not displace current neuroscience but adds a new dimension to it by suggesting a possible mechanism for the basis on which consciousness acts through the brain. Both universal consciousness as it manifests through quantum processes and neuroscience are required. Of more importance, however, this view heals science, adds meaning and purpose to life again, and links us to the primary creative force of the universe.

Wider Experiences

It is clear from the above that, however hard we look into matter as conceived of by physics and brain processes as structured by neuroanatomy and neurophysiology, we can never come across consciousness. It has already been excluded by our method. The only way to start the process of making matter conscious again is to look at those special experiences in which consciousness is primary and the experiencer gets closer to a wider conscious state, for example mystical experiences and the experiences surrounding death.

Spiritual experiences, such as the feeling of being in the presence of a power greater than oneself, are very common in the population. There are many studies of the frequency of mystical or religious experience. Glock and Starck (1965) showed that over 45% of Protestants and 43% of Roman Catholics had had 'a feeling that you are somehow in the presence of God.' Gallup surveys in the United States by Back and Bourque in 1963, 1966 and 1967, showed that 20.5%, 32%, and 44% respectively had had religious or mystical experiences and the percentage increased as the decade advanced. (Back and Bourque 1970). However, by 1978 the Princeton Religious Research Centre found, in answer to a similar question, that the positive response was down to 35%, possibly a reflection of the waning of popular interest in the mystical. In Britain, David Hay organised an NOP survey in 1976, asking a similar question, and found a similar rate of reply: about 36% gave positive responses. However, such experiences now seem to be on the increase again according to recent UK research by Hay and Heald. The incidence has risen from 46% in 1987 to 76% in 2000. The poet W.B. Yeats described the experience well:

> *My fiftieth year had come and gone*
> *I sat, a solitary man,*
> *In a crowded London shop,*
> *An open book an empty cup*
> *On the marble table-top.*
> *While on the shop and street I gazed*
> *My body of a sudden blazed*
> *And twenty minutes more or less*
> *It seemed, so great my happiness*
> *That I was blessed and could bless*

The sense of reality associated with these experiences does seem to suggest that they have a dimension beyond the physical. They cannot simply be reduced to localised neuronal firing.

Ecstatic mystical states in which the subjects describe a feeling of universal love, or become identified with some aspect of the cosmos, occur much less often. These states can occur spontaneously, but they, or fragments of them, may also occur in other circumstances, for example in the near death experience. Such alterations in mental state can also be induced by a number of quite common meditation techniques, while some hallucinogenic drugs can induce similar mental states. They can occasion-ally occur in temporal lobe epilepsy, and also in psychosis, when they are usually associated with an elevation of mood.

It therefore seems probable that the ability to experience these wide mystical states is a normal part of brain function, and indeed, there are techniques in many Eastern religions directed at inducing these wide feelings of universal love at will. Bucke, a nineteenth-century Canadian psychiatrist, was one of the first Western scientists to attempt to define mystical experience. In his book Cosmic Consciousness he examined many very deep experiences:

> Now came a period of rapture so intense that the universe stood still as if amazed at the unutterable majesty of the spec-tacle: only one in all the infinite universe. The all-caring, per-fect one, perfect wisdom, truth, love and purity: and with rapture came insight. In that same wonderful moment of what might be called supernal bliss came illumination.... What joy when I saw there was no break in the chain — not a link left out — everything in its place and time, worlds, sys-tems, all blended in one harmonious world, universal, syn-onymous with universal love.

Bucke and others have listed nine features that characterise the main elements of the mystical experience. These are:

- feelings of unity
- feelings of objectivity and reality
- transcendence of space and time
- a sense of sacredness
- deepy felt positive mood — joy, blessedness, peace, and bliss.
- containing paradox-mystical consciousness which is often felt to be true, despite a violation of Aristotelian logic.
- ineffability — language is inadequate to express the experiences
- transiency
- positive change in attitude or behaviour following the experience

Although all these features are quoted widely in the mystical literature, they are not in any way limited to rare spontaneous mystical states, but are part of normal human experience. They are also a feature of pathological experiences such as psychoses.

If mystical experience is so common, then it is logical to assume that there must be a brain mechanism which allows expression of the experience. The question then is, what mechanism? Much of the evidence we have about the brain mechanisms which mediate that state has been acquired through the study of pathologically-induced mystical experiences. Epileptic and drug states, particularly that due to ketamine, an NMDA receptor agonist, are such examples. There is also a significant literature on the neuroscience of meditation and the effectiveness and possible mechanisms of prayer. Finally, there are the beginnings of a literature on sets of unusual phenomena which occur in the 24 hours before death and which suggest a wider view of consciousness. The near death experience has also been suggested as a way for answering the question, are brain processes entirely responsible for consciousness? Is the mind/brain identity theory correct or are there situations in which mind and brain can be separated.

Over the last 20 years a major change in scientific thinking has occurred, with a recognition of spiritual values. Perhaps one of the first indicators of change has been a recognition that our thoughts and mental activity are linked to and have a powerful influence on our mental and physical health. Studies of meditation show this very clearly. The main change that occurs in meditation is the relaxation response associated with a decrease in cortisol, a decrease in blood pressure and a reduction in the galvanic skin response. Changes in the EEG have also been described , which in long term meditators are present between the meditation sessions. Newburg et al (2001) published a SPECT scan study in a group of eight meditators and found changes in the frontal and parietal cortex that were closely related to the positive mental states and changes in body image that are described during meditation.

A study by Aftanas and Golocheikine (2001) has examined the EEG during Sahaja yoga meditation. They were particularly interested in the correlation of the electrical activity with feelings of bliss and the difference between long and short term meditators. In long term meditators they found a correlation between brain states and feelings of bliss and a negative correlation with the appearance of thoughts. Meditation has been used for a number of years in clinical practice and has been found to be generally helpful in those clinical conditions where high arousal and anxiety are part of the pathology. Kabat Zinn (1992) found a significant improvement in patients with general anxiety and panic disorders who were treated with counselling and mindfulness meditation.

Even more interesting is the power of compassion and positive thinking. McClellan (1998) measured the level of salivary IgA, an immunoglobulin, in a group of students, 70 of whom watched a film of Mother Teresa healing while 62, the controls, viewed a film about the triumph of the Axis Powers in World War II. He found a significant increase ($p=0.025$) in salivary IgA following the watching of the film of Mother Teresa healing with compassion compared to the control group which he suggested indicated an up-regulation of the immune system. A second study measured the effects of the visualisation of Mother Teresa healing. This also produced an increase in immunoglobulin , and an up-regulation of the immune system.

One of the early indications of the general acknowledgement of the relevance of spirituality was a paper on prayer (Byrd *et al.* 1983) which showed that patients in a coronary care unit who were prayed for had shorter in-patient stays, fewer complications and a reduced drug usage.

In our Judaeo-Christian society prayer has always been recognised as a way of ameliorating distress. Let us start by looking at the way in which the power of prayer was regarded over a century ago. This is a story told by a country lawyer in the mid-west of the US around 1880. There had been a prolonged drought in the area, so long and so severe that the little farming community in which he lived was facing grave difficulties. Without rain, and rain soon, the harvest would be ruined. All the ministers of the various churches in the community put their heads together, and it was resolved that at midday on a certain day, the whole community would pray for rain. On the day in question, at midday, work in the town came to a standstill and as a body, the town prayed for rain. After a couple of hours, the sky clouded over and the rain came. It came in bucketsful, accompanied by thunder and by lightning. Unfortunately the lightning struck and completely destroyed a barn, which was, again unfortunately, not insured. The owner of the barn, a sceptical town worthy, decided to sue the preacher whose idea this had been, and the lawyer was asked by the priest to represent him. The case achieved some notoriety and in fact went all the way to the US Supreme Court. Did the plaintiff, asked the judge, believe in prayer? The plaintiff started to say he did not, but realising this would be undermining his own case, finally mumbled that Yes, he did. Did he pray for rain? Yes. Did he pray also for lightning? Why, no. Then, said the judge, the lightning is an act of God and the case is dismissed!

Prayer is a widespread practice in the US: Newsweek found in a 1992 survey that nine of ten Americans prayed at least once a week. In 1994 Life magazine found the same proportion believe that God answers their prayers, while Time found in 1996 that 82% of Americans believe that prayer heals. And maybe they are right. In the last few years there have been a number of high quality double blind randomised control trials of intercessory prayer, the majority of which have been positive, and which have produced good supportive evidence that prayer is effective; numerous other studies have shown the success of spiritual medicine in general. These studies come from a wide number of fields and deal with many aspects of religious and spiritual practice. A study of over 65s by Koenig *et al.* (1999), for example, was able to show that church attenders were more likely to be alive after six years. A further study has shown that over 65 year olds who went to church at least once a week had lower levels of cancer and heart disease, suggesting an up-regulation of the immune system. Koenig *et al.* (*Handbook of Religion and Health*, 2001) showed the beneficial influence on health of having a religious or spiritual belief. A strong faith, a supportive social network, positive relationships and positive thinking up-regulate the immune system, reducing the risk of cancer and heart disease and improving general health.

The first good double blind randomised control trial of prayer is that already mentioned of Byrd *et al* (1988) carried out in a coronary care unit. The names of the active group were sent to a prayer group who were instructed to pray that those named would get better more quickly, have fewer complications etc. The results were significant. In the prayed for group there was a five fold reduction in the use of antibiotics, a three fold reduction in the occurrence of pulmonary oedema, fewer subjects required intubation and fewer subjects (though not significantly fewer) died than in the control group. This paper became the model for a number of further studies.   Another notable study is that of Cha et al (2001) from the department of obstetrics and gynaecology at Columbia Hospital New York. They carried out a prospective double blind randomised control trial on the effects of intercessory prayer on in vitro fertilisation and embryo transfer in a group of patients in Seoul, South Korea. There were three praying groups, one in Australia, one in the USA and one in Canada. The results were striking. The prayed for group showed higher implantation rates (16.3% against 8% for the control) (p=0.0005) and higher pregnancy rates (50% against 26%) (p=0.0013). The high significances suggest again that prayer is effective; and the fact that the people praying were widely separated from those they prayed for suggests that action at a distance has to be postulated and that some kind of intention to heal on the part of those praying crosses space to influence the target group. This study is thus a parapsychological study on healing and suggests the possibility of direct effects of mind beyond the brain. There is now a growing scientific recognition that mind must be non-local and that effects beyond the brain must be considered in any current theory of consciousness.

Dying and Death Bed Experiences

One of the central planks of scientific materialism is that when the brain dies, consciousness ends. So I would like to look now at the phenomena which occur when people are approaching death, and also at the near death experiences that people report in cardiac care units when their hearts have stopped. These phenomena are so meaningful to the people who have them, so widespread and so indicative of a continuation of consciousness after death that they deserve more detailed investigation than has yet been carried out, as they may help in determining whether consciousness exists beyond the brain.

Various phenomena have often been reported by carers and relatives of the dying in the 24 hours before death. In these 'approaching death' experiences the dying person may report seeing visions, usually of dead relatives or friends who seem to have come to 'take them away' — to help them through the death process. Many people reporting such visions become (or are already) lucid at the time of the experience and die soon afterwards. There is clearly a cultural component in the visions seen. Osis and Haraldsson (1997) found that in a western culture the vast majority — 70% of figures seen — were apparitions of the dead, 17% were of people still alive, and 13% were of religious figures. A similar study in India produced very high rates of religious figures specific to the religion of the person having the vision. (*At the Hour of Death*, 2nd edition 1997). Other studies indicate that 30% of those apparently seeing deathbed visions die getting out of bed as if they were trying to go somewhere. An Italian paper (Giovetti, 1999) quotes this case of the vision of a dying man. His wife reported: 'The gauze over his face moved. I ran to him. 'Adriana my dear, your mother' (who had died three years earlier) 'is helping me to break out of this disgusting body. There is so much light here, so much peace.' Giovanni found that 40% of the visions she reviewed were of this 'take-away' type.

Sometimes the dying person experiences light, and seems to drift into and out of another world. Indeed, light is one of the phenomena quite often described by carers of the dying.

> Suddenly there was the most brilliant light shining from my husband's chest, and as this light lifted upwards there was the most beautiful music and singing voices. My own chest seemed to fill with infinite joy and my heart felt as if it was lifting to join this light and music. Suddenly there was a hand on my shoulder and a nurse said 'I'm sorry love. He has just gone.' I lost sight of the light and music. I felt so bereft at being left behind.

It is interesting that light is one of the recurring themes in mystical experience. It is usually brilliant white or gold, is felt as embracing and warm, and is experienced concurrently as love. Love is intertwined with light and is experienced as universal, interpenetrating everything.

Another set of experiences which are sometimes reported are the deathbed 'coincidences', when close friends or relatives of the dying person say that they have seen, or become aware of them at the time of their death. Often they don't even know their friend or relative is ill. These experiences were well documented as long ago as 1886 in the two-volume study *Phantasms of the Living* by Gurney, Podmore and Myers.

The following account was sent to me after I was interviewed on radio about near death experiences.

> When I retired to bed I was very restless. I tossed this way and that until suddenly in the early hours my father stood by me bed. He had been ill for a long time, but there he was standing in his prime of life. He didn't speak. My restlessness ceased and I fell asleep. In the morning I knew … my father had died late the evening before and had been permitted to visit me on his way into the next life.

Death bed visions certainly contain a strong cultural component and this may lead to a scientific interpretation that they are totally culturally determined, and that it is the effect of culture in the circumstances surrounding death that leads to the appearance of the visions. Against this are the phenomena that are found in intensive care units, the intensive care psychosis, when patients are in a state of sensory deprivation, helplessness and confusion that accompany severe illness. The studies on intensive care psychosis suggest that the phenomena are mainly paranoid and accusatory. They tend to be confusional and are certainly very different from the clarity of the death bed vision. The difference in phenomenology between these two sets of experiences suggests that they may well have a different aetiology. Further evidence that death bed visions may contain indications of a continuation of consciousness is given by those rare occurrences when the dying person 'sees' a relative whom they did not know was dead. Often the death of this relative has been kept from them by the family. For example a son who had recently died in a road traffic accident appeared in his mother's death bed vision, although she did not know he had been killed. However it must be remembered that some death bed visions are of the living, and so only a larger study will help resolve whether the dead person was simply seen by chance. The only way to approach this is through further and more detailed studies of the phenomena, correlating these with the mental state of the individual.

Death bed coincidences firmly point to the non-locality of mind and the entanglement of the dying person with the person who is visited. Although this entanglement occurs remotely, it does not necessarily point to a continuation of consciousness, as some form of telepathic contact could explain the phenomenon, but it would be consistent with separation of brain and consciousness at the time of death.

Near Death Experiences

The near death experience has now been well documented. It differs from death bed visions as it frequently occurs in people who are deeply unconscious. Its main features are a feeling of peace, reported by 82%, experience of light 72%, an out of body experience 66%; 49% go through a tunnel and 33% meet a being of light; 76% report pastoral landscapes, 38% meeting dead friends or relatives, 12% have a life review, 24% come to a barrier and can go no further, and 72% make a decision to return or have such a decision made for them. In addition, 72% report being transformed in some way by the experience.   However, these phenomena can also occur in people who are healthy, often precipitated by fear, drugs, pain etc. — in fact in randomised surveys about 10% of people in the general population report that they have had such experiences. It is difficult to examine these experiences as one can never be certain of the exact mental state at the time of the experience, nor can they can be predicted.

The best 'near-death' model is that of cardiac arrest, when 10% of patients do have the experience. In these circumstances, the heart has stopped and the patient is clinically dead — that is, they have no detectable cardiac output, no respiratory effort, and an absence of brain stem reflexes — no gag reflex and dilated, unresponsive pupils. A number of people say they leave their body during an NDE and it might be possible to test whether or not they get new information while they are out of their body, ostensibly when the heart has stopped.

For the scientific researcher, the interesting question is this: when does the NDE occur? Does it occur before or during unconsciousness caused by the heart attack, during recovery or after recovery? The onset of unconsciousness is very quick, as occurs in a faint, so the experience could not occur then. During unconsciousness all brain functions cease so neuroscience says the experience could not occur then either. During recovery from a cardiac arrest the subject is confused so the clear, lucid NDE could not occur in this confusional state. If it could be shown scientifically that the near death experience occurs during unconsciousness, as suggested by those who have survived a cardiac arrest, when all brain function has ceased and there is apparently no mechanism to mediate it, this would be highly significant, because it would suggest that consciousness can indeed exist independently of a functioning brain.

Dr. Sam Parnia and I therefore carried out a study in the coronary care unit at Southampton Hospital. 63 cardiac arrest survivors were questioned over the course of a year. 56 had no memories during the arrest and formed the control group, and seven, 11%, had memories, of which six were clearly near death experiences. These experiences were the same as are described in the literature and included the phenomena described above typical of near death experiences, for example the tunnel, the light, then meeting a barrier and returning. There were no out of body experiences in our study. There were no physiological differences between the NDE group and the control group, except that the experiencers had super-high levels of oxygen.

All the experiencers reported that their experiences occurred while they were unconscious, but the only way of testing this is to see if patients can gain new information when they are out of their body. There are many anecdotal accounts of information being obtained in this way, some of them very persuasive, but none providing the kind of proof that science requires. Our own method was to put pictures on the ceiling in the hope that they would be viewed by the experiencers. In this particular study none of those who had an NDE left their body. In another study carried out in Wales by Penny Sartori, about 8 people did leave their body, but none of them looked at the cards, which were on a monitor beside the bed. So we still don't know!

There are now other well conducted studies of NDEs during cardiac arrest which confirm the difficulty of reconciling accounts of NDEs with the accepted picture of brain function during a cardiac arrest. 'Near death experience in survivors of cardiac arrest: a prospective study in the Netherlands,' by van Lommel et al, was published in the *Lancet* in December 2001. This was a large study carried out in ten hospitals and with over 340 cases of NDEs and an 8 year follow-up. A US study by Schwamninger *et al.* on 'The NDE in cardiac arrest' was published in the *Journal of Near Death Studies*, 2002.

Bruce Greyson, who has published a similar study, has commented:

'The paradoxical occurrence of heightened awareness and logical thought processes, without subsequent amnesia, during a period of impaired cerebral perfusion, raises particularly perplexing questions for our current understanding of consciousness and its relation to brain function.

The question of when the NDE occurs should be answered when sufficient funding is available to carry out what would be a relatively cheap research project, but one with enormous potential for society. Does the NDE do anything to confirm the continuation of personal consciousness after death? People who have had the experience say that it does. But to accept this we have to step beyond the boundaries of science, which does not accept subjective experience as evidence. Only by trying to establish a new science, which does attempt to explore and to validate human subjective experience can we decide whether there is in fact meaning behind the mechanism of the NDE and perhaps even allow for the continuation of personal consciousness.

The strength of the view that there is a continuation of personal consciousness is that it allows for personal responsibility. A purely mechanistic and deterministic view of the world removes any possibility of personal transcendent values. We need a wider view to encompass the instinctive belief of many people, that we are not simply mechanical entities responsible only to our biology and to our culture, but that there is a spiritual aspect to our nature, that we are part of a greater whole, and that we carry personal responsibility for our actions.

This leads into a consideration of the ethics of the NDE. If consciousness is found to continue after normal electrical activity in the brain has ceased, this suggests that we may be fundamentally spiritual beings. Examination of the phenomena surrounding death and the dying process itself suggests that there is a process to dying, and this also requires further research.

There is now strong scientific evidence that mind has to be considered to be distributed. Non-local effects have been proved in many studies. Remote brain to brain communication has been confirmed by EEG measurements. Telepathy has been confirmed by Professor Robert Morris's Ganzfeld experiments in Edinburgh, and has been shown to occur significantly in daily life (Rupert Sheldrake's recent experiments on telephone and email telepathy, for example — see www.sheldrake.org). Prayer and healing suggest action of mind at a distance. Approaching death and near death phenomena also raise the possibility of mind beyond the brain.

Science itself suggests that we may now need a different way of looking at the brain and at consciousness — a science of spirituality. Western science has shown the beneficial effects on health of leading a spiritual life. It has shown the high degree of interconnectedness between individuals and suggested that any theory of consciousness must include mind beyond the brain. Western science has even raised the question of a continuation of consciousness beyond death.

Both transcendent experiences and the near death experience appear to give a subjective view of what lies beyond the physical, suggesting that the very structure of the world is spiritual, that consciousness is primary and unitary and that individual consciousness is part of the whole and survives death. From these experiences there appears to be a moral order in the universe with personal responsibility for all your acts and even thoughts. This indicates an underlying unity of consciousness, a level at which we are not separate from each other and, in my view, provides an ontological basis for the Golden Rule, to love your neighbour as yourself, expecting nothing in return. The underlying inner structure of the universe consists of light and love, which are traditional central attributes of the Divine, as reflected in the Persian poet Hafiz's view of love:

> I have estimated the influence of Reason upon Love and found that it is like that of a raindrop upon the ocean, which makes one little mark upon the water's face and disappears.

References

Aftanas, L.I. & Golocheikine, S.A. (2001), 'Human anterior and frontal midline theta and lower alpha reflect emotionally positive state and internalised attention: high resolution EEG investigation of meditation', *Neuroscience Letters*, **310** (1), pp. 57–60.

Back K.W. & Bourque, L.B. (1970), 'Can feelings be enumerated?', *Behav. Sci.*, **15** (6), pp. 487–496.

Byrd, R.C. (1988), 'Positive therapeutic effects of intercessory prayer in a coronary care unit population', *Southern Medical Journal*, **81**, pp. 826–9.

Cha, K.Y, Wirth, D.P. Lobo, R.A. (2001), 'Does prayer influence the success of in vitro fertilisation-embryo transfer? Report of a masked randomised trial', *J. Reprod. Med.*, **49** (9), pp. 781–787

Dennett, D.C. (1991), *Consciousness Explained*. London: Penguin.

Giovetti, P. (1999), 'Visions of the dead: Death-bed visions and near-death experiences in Italy', *Human Nature*, **1** (1) — Psychical Research.

Glock, C.Y. and Starck, R. (1965) *Religion and Society in Tension*. Chicago: Rand McNally.

Goswami, A. (1993), *The Self-Aware Universe*. New York: Simon & Schuster.

Greyson, B. (2003), 'Incidence and correlates of near-death experiences in a cardiac care unit', *General Hospital Psychiatry*, **25**, pp. 269–276

Hameroff. S., Nip, A., Porter, M., Tuszynski, J. (2002), 'Conduction pathways in microtubules, biological quantum computation, and consciousness', *Biosystems*, **64**, pp. 149–168.

Hay, D. and Hunt, K. (2000), 'Understanding the spirituality of people who dont go to church: A report on the findings of the adults' spirituality', Project at the University of Nottingham

Kabat-Zinn , J., Massion A.O., Kristeller J. *et al.* (1992) . 'Effectiveness of a meditation-based stress reduction programme in the treatment of anxiety disorders', *American Journal of Psychiatry*, **149**, pp. 936–943.

Koenig, H. *et al.* (2001), *Handbook of Religion and Health*. Oxford University Press.

Koenig, H. *et al.* (1999), 'Does religious attendance prolong survival? A six year follow up study of 3968 older adults', *Journal of Gerontology* (Medical Sciences), **54a**, pp. m370–377.

McClelland (1998), 'The effect of motivational arousal through films on salivary immunoglobulin A', *Psychology and Health*, **2**, pp. 31–52.

Nagel, T. (1974), 'What is it like to be a bat?', *Philosophical Review*, **83**, pp. 435–50

Newburg, A. et al (2001), 'The measurement of regional cerebral blood flow during the complex cognitive task of meditation: a preliminary SPECT study', *Psychiatry Research*, **106** (2), pp. 113–122.

Osis, K. and Haraldsson, E. (1997), *At the Hour of Death*, 2nd edition.

Parnia, S., Waller D., Yeates R., Fenwick P. (2002), 'A qualitative and quantitative study of the incidence, features and aetiology of near death experiences in cardiac arrest survivors', *Resuscitation*, **48**, pp. 149–156.

Radin, D. (1997), *The Conscious Universe*. Harper: San Francisco.

Schwamninger, J. (2002), 'A prospective study of near death experiences in cardiac arrest patients', *Journal of Near Death Studies*, **20** (4).

Searle, J. (1992), 'The problem of consciousness', in CIBA Foundation symposium no 174 *Experimental and Theoretical Studies of Consciousness*, Chichester: John Wiley pp 61-80.

Sperry R.W. (1987), 'Structure and significance of the consciousness revolution', *The Journal of Mind and Behaviour*, **8**, p. 12.

Van Lommel, P., Wees Vann R., Meyers V, Elferrich (2001), 'Near-death experience in survivors of cardiac arrest: a prospective study in the Netherlands', *Lancet*, **358**, pp. 2039–2045.

Velmans, M. (2000), *Understanding Consciousness*. London: Routledge, Psychology Press.

Guy Claxton

Proximal Spirituality

Why the Brains of Angels are Different from Ours

'It's Simple To Be Happy.
It's Just Hard To Be Simple'

Do you want there to be angels? And if so, in what sense do you want them to be real? Is it inconceivable to you that the world of Newton, Lavoisier and Darwin, even of Bohr and Einstein and Feynman, is the only world there is, and utterly obvious that there must be another parallel world, or perhaps many such worlds, which transcend the mundane laws of matter and biology: worlds which contain the supernatural spirits, forces and entities — angels and astral bodies, gods and morphic fields, aliens and 'the power of evil' — that are needed to underpin your deepest beliefs and most profound experience? Is 'evil' as real as 'gravity'; is Heaven almost as substantial as Harrogate? Or are the Devil and 'telepathy' merely real in the way that the Bogeyman, Cinderella and The Holy Grail are real: as powerful mythic figures and numinous symbols that point to or encapsulate vital, mysterious aspects of human existence, but without causing any worry that the Theory of Relativity has been breached?

'The heart has its reasons of which reason itself knows nothing,' said Pascal, and our pre-rational hopes and beliefs tilt the surface of our rationality, so that, without noticing, we deploy our credulity and our scepticism partially. If we want the supernatural to be real, we are readier to accept a professor's word that the study demonstrating the power of prayer was well-conducted; and readier to accuse the sceptic of a wilful refusal to accept 'the facts'. If we want the Book of Common Prayer to be poetry, and the Voice of God to be a powerful, meaningful, beautiful hallucination, then we shall challenge the professor's claim more robustly, and work harder to make the psychological explanation stick. Even Tibetan Buddhist lamas fight about whether reincarnation is to be taken literally or figuratively, so if we think we are fighting fairly, when it comes to the supernatural, we are deluded. 'May the best man win', we say, smiling, as we slip the tiny capsule of Rohypnol into the opponent's tea.

Peter Fenwick and I rig the scales so that the same weight of evidence tips them in different ways. His apparent even-handedness masks his wish that the Out of the Body experiencer really does float up to the ceiling, and can thus see the secret message scrawled in the dust on the top of the high cupboard. I want it not to be so, and will ferret out the flaw in the experiment that allows my pretheoretical commitment to escape. Neither of us is innocent, and we would be better employed meditating on what we each need to be true, and why, than on trying to establish the impersonal truth, for we shall never agree on what is sufficient proof.

Such introspection, however, is a protracted and a private business, and the chapter has to be written. So I shall distract you from my sleight of hand with the murmur of sweet reason, and tell you that I am not trying to inveigle you into *believing* that angels have brains, and not wings, but simply to join me in *exploring* how far the world of cognitive and neuroscience can be stretched, without being broken, in the direction of accounting for some, at least, of the more puzzling aspects of human mystical and spiritual experience. The supernatural believer will sit with a sad or cynical smile, knowing that my efforts are doomed to fail — that science will snap before it gets to the *really* interesting stuff. She knows that the message on the top of the cupboard *will* be read, and that its reading will floor me. And the materialist will cheer me on. At the end of the day there is no such thing as an open mind. The mind is not that kind of thing. An 'open mind' is ultimately an oxymoron (better, I suppose, than a 'closed mind', which is just a moron). So be it.

For the sceptic, 'Ultimate Reality' itself is also a rich fiction, by the way: a hypothetical realm, a creation of human imagination no more or less lovely and deep than the Happy-Hunt-ing-Ground is to a Native American, or Never-Never-Land to generations of children. As children grow up, Santa Claus stops being real in a literal, childish sense, and becomes a benign and beautiful symbol of just reward and generosity; and as we grow up, so Ultimate Reality stops being a hidden domain and becomes a reminder of the sense of depth and mystery that lies at the heart of all meticulously observed human experience. Plato's cave is a misleading metaphor, for it is not just unusual in prac-tice, but impossible in principle, for us to turn round from the shadow-play upon the wall in front of us, and see 'directly', 'truly', what is 'really' going on behind our backs. Our physiol-ogy is insensitive to all but a fraction of the electromagnetic hum of the world, and no amount of grace or meditation will make it otherwise. Our eyes are built to exaggerate discontinuities, and even — as many illusions reveal — to invent edges and contours that are not 'there' in any physical sense at all. I am blind to my own 'blind spot', not just through laziness but inherently. Our brains are designed, for good reasons, to detect changes in the world that occur at certain paces, and not to notice others that are significantly faster or slower: the turning of the tide, or the flicker of a fluorescent tube. There is so much that we simply cannot sense — and would perish if we did — that Ultimate Reality is doomed always to remain an idea.

And if my 'meat' ineluctably narrows and distorts my organis-mic sensitivity to what is out there, how much more partial and untrustworthy do I know my consciousness to be. So much more is going on in me than I am ever aware of, and if suddenly the blinds were to be drawn aside, and I were to witness the full complexity of my insulin receptors taking soundings of the blood, and the detailed pattern of pressure that the floor makes on the soles of my feet, and the thousand preoccupations that normally hum inaudibly at the back of my mind, and ... and ... and ... I should drown in data. 'Too much information' is anath-ema, and that refers not just to emails but to the moment-to-moment 'reality' of my own body and brain. Even the Ultimate Reality of my own process is unobservable, and again, for good reason.

The experience and beliefs that I have picked up from my fam-ily and culture — mostly unwittingly — inevitably skew and col-our my perception. From those around me I have learnt what to notice, what to value, what to fear, and what to worship, and I would not be able to function as a member of my society if I were to strip away these enculturated habits. More deeply than I can possibly ever articulate, I know what it is to be a man, to be English, to relate to the world in a 'male, English' kind of way. I have preferences and expectations, hopes and fears, and it could not be any other way. Even after their enlightenment, all Buddhas still prefer tea to coffee (or vice versa), still feel a surge of nausea at the idea of doing a bungee-jump, still wear the habits of their nation as they shop, eat and stow the futon or shake out the duvet. Religious experience does not involve a dropping of all conditioning, and standing naked in some kind of generic humanity. Personality does not disappear; one simply stands in a different relationship to it.

Thus it is childish to take one's conscious experience at face value: to assert that because I saw it, heard it, smelled it, dreamed it, it must be 'real'. 'Because I heard my name, someone must have said it' is a false belief, for I know how often someone in the noisy street shouted 'Bye!' to a friend, and I heard 'Guy'. I remember the time I swore that I had given the girl a twenty pound note, when she only gave me change for ten, and then found the note at the back of my wallet — so I know that 'because I remembered it, it must have happened' is an article of faith rather than a trustworthy fact. How often I have known that it was you who 'started it', and then had to swallow my words and admit that you were just a projection of my own bad day. 'He makes me mad' — as if there were a direct causal link between what he does and how I feel — is often a lie, and always open to dispute.

The evidence from cognitive neuroscience further undermines any hubristic tendency to claim physical, independent reality for the things we experience, and demonstrates just how much of our perception is imputed rather than extracted. Hungry people see coins as larger than sated people do. Autistic people see the world in more accurate detail than the rest of us. We know how easy it is to implant false memories. We know that the wrap-around visual world in which we think we live is actually highly patchy and schematic, and that we are simply oblivious to most of the 'holes' that it contains. It's an Emmental world that we think is made of Cheddar. We can even be confused about whether an experience is a new perception or a retrieved memory. (In a recognition test, make some of the candidate stimuli slightly fuzzy, and people will say that the clear ones, even if 'new', are memories, presumably because they infer, from the slightly faster processing that the sharp stimuli allow, that they must be more 'familiar'.) Activity in the right anterior cingulate area of the brain determined whether we experience something as a perception or as imagination. 'We see things not as they are, but as we are', said Kurt Koffka. Hallucinations and misinterpretations are not the provinces of psychotics and magicians; they belong to us all. As Buddhist scholar Alan Wallace says, the evidence, unwelcome though it may be, is that we think we are looking at the world through a telescope, and it turns out to be a kaleidoscope.

And *a fortiori*, in the territory of unusual or mystical experience, the skeptic says we have to be especially cautious, tentative and humble, and remain wary of the tendency to reach for a supernatural explanation for what we do not immediately understand — and then to swear by it. The experience of being abducted by aliens is real enough, but is not evidence at all that aliens are 'real' in the literal, childish sense. The experience of floating above one's own body is powerful and strange, but that does not mean that one 'really' did. You may have done — but the sceptic remembers how often she has been wrong before about more footling matters, and reserves her judgement. If those who forget history are condemned to repeat it, it is those who know no psychology, and are unused to being quizzical about their own process, who are most likely condemned to mistake the unusual products of their own brains for actual events in the external world. Most people, these days, are reluctant to claim that they really flew in their dream last night, and to accept that its so-solid reality was a figment of their own brains; yet they may still swear that the goosebumps in the castle corridor must have been due to a 'presence' whose reality is not in doubt.

Perhaps the most ubiquitous 'category error' of this kind, says the confirmed sceptic, is the belief that behind their thinking lurks a thinker; behind their action an actor; behind their experience an experiencer, behind their decision a decider: that there is a ghost in their own machine more real and incontestable than the one in the mediaeval passageway. Descartes's *cogito ergo sum* is not a ground- breaking discovery but a highly dubious imputation. The fact is, we are aware of thinking, of noticing, of deciding going on, and then we theorise these, as a society, in terms of a shadowy inner agent who is pulling the strings. The theory then seeps into our minds, so that the ghostly controller becomes almost tangible — yet, if we are to be consistent, we may not take that sense as *prima facie* evidence for the real existence, or the real efficacy, of the ghost. The fact that so many people share in this interpretation — what Charles Tart refers to as the 'consensual trance' — makes it harder to doubt, but no more likely to be true.

So the skeptic's question is: which, and how many, of our imputations, interpretations and projections are benign, or even useful, and which cause us trouble; and which are hard-wired, and which are mutable. The skeptic's spirituality is a rather dull and low-tech one. It cannot involve an encounter with Ultimate Reality, for she knows that that is an impossibility. And it has little time for the technicolour fantasies of 'visions', for these are magnificent dreams, bursting with significance, but dreams none the less. Collecting and pressing them in memory, like wild flowers, is beside the point. In Zen Buddhism such colourful experiences are treated as *makyo*, distractions, to be dispassionately noted and allowed to float away. When an excited meditator told Maharishi Mahesh Yogi that she had just had a dramatic vision of Jesus, Maharishi advised her, if it happened again, to 'shake his hand, say hello ... and come back to the mantra'.

For some people, spirituality is about deep beliefs and finding conscious answers to the Big Questions of life and death and meaning. For others, it is about seeking and collecting these rare, exotic blooms of experience, and interpreting them as compelling evidence of Another World — preferably a World That Science Can't Explain. But for our poor down-beat skeptic, spirituality is just a matter of trying to clean the gizmo through which she experiences the world, so that she can tell what are real stars, what are specks of dirt that have got stuck to the lens, and what are intrinsic limitations of the instrument — simply reflections of the kind of 'telescope' she happens to be. Relative cleansing is the only game in town, and some incremental enhancement of the quality of their lives the only reward. It is not dramatic episodes that matter so much as a sustainable change in the quality of the picture — regardless of what the story-line happens to be. As one 93-year-old lady, one of my down-beat skeptics, wrote:

If I had my life to live over, I'd try to make more mistakes next time. I would relax. I would be sillier than I have been this trip. I know of very few things I would take seriously. I would take more chances. I would take more trips. I would climb more mountains, swim more rivers and watch more sunsets. I would eat more ice cream, and less beans. I would have more actual troubles, and fewer imaginary ones. You see, I am one of those people who live...sanely and sensibly, hour after hour, day after day. Oh, I have had my moments and, if I had to do it over again, I'd have more of them. In fact, I'd try to have nothing else. Just moments, one after the other. If I had my life to live over, I would start bare-footed earlier in the spring and stay that way later in the fall. I would play hooky more. I would ride on more merry-go-rounds. I'd pick more daisies.

She doesn't want philosophical answers to 'what it's all about', or even visions of the Virgin Mary; she wants to be clearer, in the midst of each moment, about what matters and what does not. She wants to get rid of some confusions about what to cherish and what to take seriously. She regrets having sacrificed her vitality for propriety. She wants to feel more, trust more, and worry less. She wants a more vivid picture on the screen of her own consciousness. She wants a better-quality, high-fidelity gizmo through which to experience life. You might call hers a 'proximal spirituality', concerned with what is intimate and near-at-hand; not a 'distal spirituality' that is grander, safer and more remote. For her, as for the Buddha, the beautiful pristine lotus has its roots in the mud of everyday life. It does not float immaculately in space. In proximal spirituality, God manifests in your attitude to ice cream (and whether you laugh when you drop it, or which stranger you buy one for, on a whim) more than He does in your musings about the afterlife, or your stories about how you once felt the *kundalini* rising, or had an inexplicable premonition about what was going to happen to your Auntie Sue.

Given that the low-key skeptic accepts that she cannot ever know what is really going on, in what ways might the quality of conscious reception nevertheless be improved? Here are some possibilities. It might become brighter and more vivid. Colours might seem *richer*, more luminous perhaps; sounds and smells stronger. Experience might be more *detailed*: you might hear the

subtle, fleeting twists of just this moment of birdsong; notice the
veins on a leaf or on the cheeks of the old man opposite you in the
train. You might feel the tingling in your toes, the throbbing in
your back, the warmth in your mouth, more keenly — for no
good reason. Perception might become *broader* and more inclusive, maybe hearing the sounds as a symphony scored for wind,
bird, keyboard and truck, rather than as a background mush or a
series of solos. You might see the unique *patterns* that uncon-
nected objects make, and enjoy the sudden burst of shadows as
the sun comes out. Perhaps part of this shift is that perception
proceeds *beyond recognition*, so that more transience and unique-
ness makes it through into consciousness, rather than just bald
exemplars of known (and 'relevant') categories. ('The eye of
desire dirties and distorts', said Herman Hesse). Visual perception can become more *wide-angle*, so that you notice and include
what is happening on the periphery of vision as well as what is
occupying centre stage. And experience may become more inclusive in the sense of incorporating awareness of the *gaps or holes*
between mental events — hearing the silence that may surround
the sounds, for example. In general, it is possible to reduce the
extent to which *attention* is focused and constrained by *intention*
— to shift from 'looking for' to 'looking at'. And it is also possible
to increase the *flexibility and fluidity* with which attention moves
between these different modes, allowing awareness to relax into
a more open, receptive attitude between bouts of more focused
searching and identification.

In developing a clearer feeling for the 'gaps' between
thoughts, for example, there may also emerge a sense of mental
contents *emerging* into consciousness, and dying away, as if
thoughts, feelings and sensations were white-capped waves
emerging out of a dark ocean, travelling for a while, and then
sinking back into the invisible body of the sea again. In this way,
a stronger feeling of the *unconscious* may develop — not as some
kind of locked ward of wild impulses and hurt memories, but as
the continual, inscrutable backdrop to consciousness: the 'wings'
that surround the well-lit central theatre of the conscious mind.
And along with that realisation may go a greater sense of the
mind's *autonomy* — that thoughts are not made by consciousness
but occur to it, and so must be 'produced' elsewhere, behind the
scenes, in an area of the mind to which 'I', *qua* conscious controller and arbiter, actual have no access. And from this gathering

realisation, two feelings can arise: *panic*, that one is not in control as much as one thought; and *delight*, at the spontaneous magic-show of thoughts, plans, memories, fantasies, feelings and sensations that is continually erupting into consciousness apparently of itself.

Thus a reappraisal of the background sense of the *instigatory self* may begin to occur, as the sense of self has to expand to include areas of the psyche that are less predictable and more mysterious. And with greater relaxation and loosening of control, it becomes possible to admit into consciousness aspects of *experience that had previously been excluded*, on the grounds that they did not 'fit' with the Cartesian idea that conscious reason was the central control-room of the mind and body. You become able to notice just how often the presumed association between thought, especially intentions, and action, is breached or compromised. You see how often you do intelligent and complicated things without thinking about them at all. And you see how often you decide to do things and then don't — or do them later, or in an entirely different way. *Thought and action* turn out, on closer inspection, to be much more loosely coupled that you had supposed — and than the conventional picture of the self requires them to be. And in dropping that requirement, the unfolding mental movie is freed to be much more *bizarre and playful*, while your attitude to this play becomes quieter and more dispassionate. Instead of fearing you are going mad, you are entertained by the surreal drama. You may indeed experience an increase in the frequency and power of 'visions' and 'voices', but they are treated simply as the amusing flotsam of the unconscious, capable of earnest interpretation if you wish, but otherwise not to be concerned about, and certainly not to be turned into 'realities' in the childish sense.

As attention becomes more *meticulous*, and at the same time calmer and less reactive, so it is possible to 'catch yourself' entering into an habitual train of thought or judgement, and hit the 'escape' key. Mischievous impulses, previously irresistible, become more likely to be nipped in the bud before they can do damage either to one's relationships with others, or to oneself. Research shows, for example, that people prone to episodes of depression can benefit from 'mindfulness training' that cultivates these shifts in mental attitude, so that, instead of 'buying into' a self-critical and self-destructive train of thought, when something goes wrong, they are able to stand outside it, observe it, and so neutralise its tendency to induce despair and self-condemnation. To put it another way, a note of affectionate scepticism enters into consciousness, in which one puts a small *ironic distance* between oneself and the mental content. Without becoming lost in a post-modern spiral of deconstruction, one begins to be able to stand outside one's own belief systems rather than being trapped within them, and thus to take a perspective on events that is less fiercely partisan or righteous, and thus more capable of *equanimity* and even of *compassion*. Decision-making is more likely to take a 'big picture' view, in which it is possible to include different points of view and values, rather than anxiously pursuing self-interest at the expense of others. Consciousness develops a calmer and kinder quality. Equanimity replaces anxiety as the dominant aura of conscious awareness. Emotions are viewed as an essential part of mental life, and not as problems to be solved or enemies to be overcome; yet, at the same time, the need to become angry or nervous, defensive or embarrassed, frustrated or jealous, hurt or acquisitive, judgemental or impatient seems to lessen.

Proximal spirituality values these kinds of shift in the quality of experience, and is centrally concerned with contemplative methods that promote their development. But how are such shifts, and the processes that give rise to them, to be accounted for? In distal spirituality, one might be inclined to invoke some 'higher powers', such as 'the grace of God' or 'the collective unconscious', to provide a pseudo-explanation. But our down-beat sceptic looks first to see if an account can be given that does not do violence to the laws of nature. She is a firm believer in Occam's Razor: the principle that theoretical entities should be invented, and established frames of reference overturned, only when it is absolutely necessary to do so. (Or she may prefer Einstein's formulation of the same idea: 'We should strive to make things as simple as possible — but not more so.') So the question is: do angels have brains? Can we account for the development of more 'angelic' qualities of mind — compassion, equanimity, generosity, lack of projection, vividness of experience — in terms of changes in the physical brain? Let us try as hard as we can — and then, if the effort fails, the theists and occultists will have been truly vindicated.

There have been two general approaches to the brain. One has a long history and a recent revival. Its working assumption is that every interesting human faculty or experience has a location in the brain where it happens, and that we can establish this location empirically. The phrenologists thought that the development of these cerebral centres of operation was reflected in the contours of the skull. Mostly it has turned out not to be so simple — though there are still claims that a person's intelligence is related to the size of their cranium. Their successors, the so-called 'neo-phrenologists', use more sophisticated neuroimaging techniques — PET, EEG, SPECT, MEG and fMRI, for example — to show which bits of the brains light up when we are doing different things. Though it has scored some conspicuous successes, the general validity of the approach is undercut by the repeated demonstration that the brain functions as a system, and not as a collection of individual little centres of intelligence or cognition. So the second approach has developed, in which researchers have tried harder to get a feel for how this intricate, integrated system might be organised. (John McCrone's book *Going Inside: A Tour Round a Single Moment of Consciousness*, gives a wonderfully lucid introduction to this approach.) It is this second, more systemic approach that I shall draw on, albeit curtly and clumsily, here.

Imagine the brain, as many neural network theorists do, as a contoured landscape, through the valleys and hollows of which flow rivers of neural activation. The 'low points' in this landscape reflect those concepts or habits that are most probable or most frequently occurring, so there is an energetic sense of activity flowing 'downhill'. As the learning brain distils useful concepts, skills and scenarios out of the recurring patterns and contingencies of experience, so the landscape is eroded, and the network of channels becomes more complex and definite. However, overlaid on this developing structural landscape of habits and associations is a much more evanescent set of 'forces' to do with recency and expectation, and desire and mood. We are set to see what is 'reasonable' not just *a priori*, but on the basis of what has just happened — an adult polar bear is rare, in my experience, but if I open my eyes and see a polar bear cub, the likelihood of the adult bear suddenly jumps — and in terms of what we want or fear to see, and what mood we are in. The landscape, if you like, is inflated and deformed by these continually shifting forces: more 'bouncy castle' than Lake District (if you can imagine them on the same scale!). Patterns of activation move through this conceptual brain-scape by finding a 'path of least resistance' that satisfies a whole complicated force-field of simultaneous constraints of need, fear and expectation. We have learned a web of expectations about what goes with what, what follows what, and so on — but still, sometimes, as George Bernard Shaw might have said of someone embarking on a fourth marriage, hope triumphs over experience; and that triumph has its underpinnings in the subtle intuitive computations of the brain.

On this picture, our goals, values, interests and fears act as pneumatic or hydraulic forces that tilt and sculpt the functional landscape of the brain, priming us to perceive and respond in ways that promote the current goal or neutralise the anticipated threat. We are sensitised to what seems 'relevant' to our current purposes (whether consciously conceived or not), and also, by the same token, desensitised to things that are judged irrelevant or anathema. Through deploying inhibition, as well as excitation, in the brain, we are able to both potentiate and attenuate: to see what we expect, and also to remain strategically oblivious to that which is too odd or inconvenient. And these tactical deformations of the brainscape take energy. The pumps and pistons that subserve motivation draw on the same limited well of mental resource — the neural activation — that also feeds the manifold streams of consciousness. As energy is drawn from this pool to deform the elastic contours of the brain-scape, so there is potentially less left to flow along the channels that have been exaggerated or bent.

Simple animals have a small set of inbuilt goals: to find periodic nourishment and a secure habitat; to mate, and protect their young for a while; to flee from or fend off the unwanted attentions of predators; to rest and recuperate; and so on. Their motives tilt their brains in relatively straightforward ways. But we humans develop incredibly complicated portfolios of interests and anxieties. To the basic need for security, we add the desire for affection and friendship, and thus turn indifference or isolation into things to be feared — unwelcome possibilities to be anticipated and neutralised. The basic need for sex and procreation is culturally encouraged, for many of us, to proliferate into a lifelong concern with appearance and 'being attractive', so that a liver spot or a wrinkle becomes transformed into an enemy.

And so on. We even espouse goals that are, in the nature of things, incompatible. We want to be popular *and* we want to be honest, and we cannot always be both. We want to be kind *and* we want to be top, and something has to give. We want to look young, and we want the leisure and self-acceptance that may come with age — so do we do our earnest, boring 20 minutes on the wretched treadmill in the spare room, or not?

As forms of desire and aversion escalate, so the brainscape becomes not just temporarily biased but permanently convoluted and contorted, so that the stream of consciousness finds a resting point of satisfaction with increasing rarity, and moment of mental quiet become few and far between. Even at night the damn thing won't shut off, because the problems which the brain has set itself, and which it is programmed to pursue, have become irresolvable. And worse, the energy needed to hold all these preoccupations in place becomes a greater and greater proportion of the total activation that is available, so that the streams of consciousness become thinner and thinner. In a paroxysm of anxiety and self-consciousness, the stream may dry up altogether, and we are left — for an instant, for a week — mentally blocked, emotionally frozen, and even physically paralysed. If self-interest conflicts with care, or delight with decorum, then more of the dwindling reservoir of resource has to be committed to deal with the glitch — by blocking the care (and then feeling guilty) or blocking the self-interest (and then feeling resentful), and then having to suppress these feelings (which undermine my self-esteem) … and so on, relentlessly and exhaustingly.

The conditioned 'self', we might say, is none other than this convoluted field of irreconcilable forces, overlaid on the workings of the brain, and altering its *modus operandi*, like a computer virus. Just as a digital computer with a serial architecture can be programmed to emulate a neural network that is massively parallel, so development can install in the human brain a program that, when it is running, makes it *look* as if its proprietor were fundamentally selfish, or tense, or greedy, or concerned about what the neighbours might say. When the Self System program is switched on, a massive expenditure of energy is required to run all the neural machinery that keeps us keyed up for what counts as profit or loss, nice or nasty, good or bad, mine or yours.

In a moment of grace, somehow or other, the brain hits the Quit button on this programme. It can happen in an instant of extreme danger, when the brain's resources are so completely stretched that there is just not enough juice left to run the Self System, and the lights of the ego flicker and go out. Such an existential power cut can happen as a result of exhaustion and sustained stress, for the same reason. And it can happen, as it does to the many thousand 'inglorious Wordsworths' who have experienced such grace, for no apparent reason at all. But when it does, as Don Juan said in one of the Carlos Castaneda books, an enormous weight of triviality drops away. The charged tangle of self-concern goes quiet, and with it, the familiar background throb of striving and insecurity. The 'bare necessities' re-emerge with clarity and force. 'What is the secret of enlightenment, Master', said the Zen student. 'Have you had your breakfast?' replied the Master. The student nodded. 'Well, go and wash your bowl.'

One of these basic motives, inherent in all social animals, and in humans most of all, is care for others, for truly each one's destiny depends on the good of all, and their identity too. We have compassionate and generous genes, as well as selfish ones. Yet the urge to be kind is often the root motive that is sacrificed when life gets complicated, resources are scarce, and you are late for a meeting. No surprise then that, when the Self System goes quiet, kindness and consideration come to the fore, as naturally as the stars come out when the sun goes down. And no surprise, either, that the energy which had been bound up in holding the brainscape warped and tilted is suddenly freed to flood into sensation and action, giving rise to a power-surge of brightness and physical energy; nor that the brainscape springs effortlessly back into shape, and the emotional world calms down and rights itself. Nor, indeed, that the awesome problem-solving — and problem-finding — capacity of the brain, no longer captured by the need to square the circle of anxious self-regard, howls with glee, and explodes with innocent creativity — as the mystics have always maintained.

How to Quit the Self System programme on purpose, rather than leaving it to chance? Certain drugs will do it — though they have yet to synthesise one that lasts for ever and has no side-effects. Mostly that solution turns out to cost more than it saves, and one of the costs is an addiction, psychological, physical or both, to a 'high' or a holiday. But High, however holy, is where you come down from, and spiritual holidays are where you have to come back from, and 'normal reception' can feel all the more intolerable on your return.

You can believe in Heaven, where fear fades and peace reigns — but not quite yet, and you can never be quite sure that virtue will really be rewarded, and that Happily — Ever — After will be all that it has been cracked up to be. (What if Heaven turns out to be *boring*, without all the hassle?) Or you can fascinate yourself with a surrogate world here and now, the parallel universe in which the limitations of time and space are transcended (and you can float up and read that message), and the frustrations of not knowing what will happen next, and of always feeling somewhat unprepared, are alleviated through telepathy and precognition. But such powers always seem to hover tantalisingly out of reach, clairvoyance remains stubbornly unreliable, and telekinesis has yet to save us air fares. This version of distal spirituality can turn out to be a counterfeit spirituality, founded on self-deception and wishful thinking, if you are not careful.

Or you can practise that almost-impossible trick — almost, but not quite — of pulling back a little and trying to catch sight of the eye through which you have been seeing; of trying to shift those biases and beliefs that have been embedded in the *process* whereby you have been experiencing into the realm of content, where they can become *objects* of experience. The effort is to catch what's there — a sound, a thought, a physical sensation — *before* the habitual conceptualisations and valencies have been added in; to move the cursor of consciousness towards the simple, unjudged state of things. Then, a split-second later, when the urge to mix in the desire and the aversion comes, it has a looser hold. This cleansing of the doors of perception is not easily achieved, nor is it usually achieved all in one fell swoop. Rather, aspects of the Self System — the embarrassment at not having anything to say, the involuntary spasm of anxiety when your joke falls flat — become partially and fleetingly visible, and therefore questionable. Sometimes, if you are lucky, the whole edifice of self-concern drops away, and the world for a while looks so lucid and lovely that you feel maybe it's not so hard to be simple, after all. More commonly, you gradually learn to live on that leading edge of experience, where sometimes you are able to pause, in the heat of the moment, and ask: 'Who says I need to need this, and fear that?', and sometimes, with a chuckle, you feel a sliver of liberation, and maybe a shiver too. That's the low-tech skeptic's modest way. That's proximal spirituality. Jesus and Buddha were proximal spirits. Jesus sadly concluded that 'except they see signs and wonders they will not believe', so he did a bit of magic in a good cause. And Buddha refused to engage with the distal stuff at all. If it was good enough for Buddha and Jesus, it's good enough for me.

References

John McCrone (1999), *Going Inside: A Tour Round a Single Moment of Consciousness*. London: Faber and Faber.

David Fontana

Science, Religion & Psychology
The Case for the Transpersonal

Introduction

Science, religion (which I take to mean a broad belief in a non-physical, spiritual dimension to existence rather than adherence to any narrow set of doctrines) and psychology may seem strange bedfellows. Is there any common ground between science and spirituality, or are the two doomed to remain at variance, with science increasingly winning the debate on the origins and nature of human existence? And does psychology have any place for non-material entities such as soul and spirit?

Before attempting to answer these questions we need to look briefly at the central reasons for the mixture of indifference and antagonism generally shown by science and by psychology towards religion, starting with science. For present purposes I mean by 'science' the prevailing philosophy and methodology that inform and determine the thoughts and practices of the most influential scientists, and that help to give science its image and status in the Western world. Why should this orthodox science, if we may call it that, appear to be so at variance with religion?

There are of course many reasons, some of them historical, but central to them is the contention that whereas religion is sustained only by faith, science eschews faith in preference for data and hard fact. Leaving aside the question of whether or not religion relies only upon faith rather than upon the direct lived experience of men and women over the centuries, let us look at the second part of this contention, namely that science deals only with data and fact.

It is indeed one of the great myths of our time (not, in fairness, propagated by all scientists themselves) that science is a precise and accurate exercise that deals in real hard facts about the world. As such, the myth has it, science is the only undertaking that can answer fundamental questions about the universe and our place in it. Unlike some myths that provide us with insights into underlying truths, this myth is dangerously misleading, and responsible not only for some of the existential problems faced by the individual seeking to find meaning and purpose in life, but for some of the destruction that we have inflicted upon our planet and upon the many species who share it with us and have equal rights to be here.

Far from being grounded always upon hard fact, science, like religion, depends in reality to a measurable degree upon faith (Wallace 2003). The difference between science and religion in this respect is that whereas most religions make clear statements of their articles of faith, science introduces its own as if they are self-evident facts. Wallace gives us a number of examples of these scientific assumptions masquerading as eternal truths. One of the most fundamental is that:

> The universe, as it exists apart from human perceptions and conceptions, can be known by means of scientific methods; although the world exists independently of our concepts, its components and laws can be grasped by concepts ...

Another is that, in spite of the fact that science has repeatedly been forced to abandon its earlier theories in the face of new discoveries that demonstrate their unreliability, it persists in the pretence that 'it is progressing steadily towards a correct representation of the universe as it is'. Wallace goes on to make it clear that these various acts of scientific faith still persist in spite of the fact that:

> For generations the notion that scientific theories represent objective, independent physical reality has been seriously challenged by philosophers of science. Indeed, there are few [such philosophers] today who adhere to such straightforward scientific realism.

Among the many problems with the realist position identified by Wallace is the fact that multiple, mutually incompatible scientific theories that account equally well for a given body of experimental evidence are often advanced by leading scientists. Despite the impression given by science that objective reality allows scientists to screen out false hypotheses leaving only the one correct theory in place, the truth of the matter is that the choice between multiple conflicting theories is often made on the basis of purely human factors, which may have much to do with vested interests such as the relative prestige and status of the scientists concerned and — increasingly in modern times — with the financial backing that they are able to command.

Wallace also draws attention to the fact that some of the supposed 'realities' discovered by scientists may be nothing of the kind. For example, the standard procedure in atomic physics (as in much of science) is to approach a body of new evidence with a specific theory in mind. The theory may prove effective in accounting for much of the evidence, but elements may be found that fail to fit in. If so, the supporters of the theory may propose that they would fit in if a new physical entity, a new particle with attributes determined by the evidence, existed. The conclusion is then reached that the proposed new particle does in fact exist, and it is given further objective 'reality' by being graced with a fancy name of some kind. As Wallace, himself a physicist with impressive credentials, points out, even Max Planck's famous idea of 'quanta' of energy, a theory that underlies modern quantum physics, is only a proposition rather than a demonstrable fact. The notion of 'quanta' provides us with an opportunity for explaining diverse and apparently incompatible microphysical phenomena, but more recently Boyer (1985) has outlined a concept which demonstrates that we can explain such phenomena without recourse to the notion of 'quanta'. In the world of microphysics, it has in fact never proved possible to make a direct observation of subatomic entities. We simply infer their existence from circumstantial evidence (e.g. the macroscopic effects — such as traces in a cloud chamber — produced when they interact with certain measuring devices). In effect, we create a hypothetical concept which has an 'as if' reality, but this is not the same as demonstrating physical realities. Physics has never been able to demonstrate that its theoretical concepts uniquely account for the experimental facts, hence the multiple incompatible theories to which we have already drawn attention and which in their various ways can each be used to account for the same phenomenon. Even mathematical 'laws' are only attempts at symbolic representations of objective reality, dependent in large measure upon the systems of measurement that we decide to use.

There is nothing intrinsically wrong with these scientific methods. Problems arise because they are presented to the public (by the media, by our educational system from school through to university, and by many high-profile scientists themselves) as if they yield accurate data about the true nature of reality. This propensity to misrepresent the scientific endeavour as an engine for discovering precise facts about the universe can actively impede the quest for truth because it prevents us from recognising the extent to which our understanding of these facts is a product of our own patterns of thinking rather than of a direct encounter with something intrinsic to that universe itself. Crucially, it prevents us from recognising the many things that science cannot do and that it is not equipped to do, things that fall within the province of other endeavours, of which religion claims to be one.

Psychology and Religion

Turning from science to psychology, we see that psychology also relies to a significant degree upon the same brand of faith that we have seen at work in science. By basing its methodology upon that used in the physical sciences, psychology shows the same readiness to accept that this methodology is a precise and accurate way of discovering real facts about the world, and that it can be applied to humans just as it can be applied to material objects. Psychology thus proceeds on the assumption that humans are nothing more than their physical bodies and brains, or that if they are more than this, then no reliable methods exist for exploring this other side of their nature. Such an assumption ignores the way in which people experience themselves and the descriptions they give of their inner lives. In spite of its emphasis upon the importance of observing behaviour, it also ignores the fact that these inner lives have given birth to many of the finest feats of human creativity. In response to religious conviction and inspiration the inner life has produced some of the most sublime music, some of the finest paintings and sculptures, some of the greatest architecture, some of the most moving poetry, some of the most illuminating moral philosophies, and some of the highest ideals known to humankind. Its readiness to dismiss this vast area of human behaviour as irrelevant to its concerns has been detrimental not only to our understanding of this behaviour but to psychology's claim to be a comprehensive discipline of human experience (Fontana 2003).

In common with the rest of science, psychology has also repeatedly been forced to abandon its earlier theories in the face of new discoveries that demonstrate their unreliability. In fact there is no single theory within psychology that has stood the test of time without extensive (and continuing) modification. And just as scientific realism has proved to be a mistaken notion, so has psychological realism, for like science psychology has also spawned multiple, mutually incompatible theories that account equally well for a given body of data, and like science the choices made between these theories are often formed on the basis of human factors rather than on the basis of hard evidence. Psychology has in fact suffered more than the rest of science at the hands of that fickle judge, fashion. Theories and explanatory models come and go in psychology with monotonous regularity, and today's leading authorities within the subject all too often become eclipsed and forgotten.

Finally, as with the rest of science, some of the supposed 'realities' discovered by psychologists are simply the products of our particular way of handling the data. For example 'intelligence', which for many years seemed to have attained the status of some objective organ of the body, as real as an arm or a leg, is in fact only a form of behaviour produced in response to certain questions asked by the intelligence test. The concept of intelligence is a useful one, and the scores people obtain on intelligence tests do give us an indication of how successful they are at solving problems in real life, but taken too literally it is also a misleading one. There are very many different kinds of problems in the real world. A person who can solve the linear, analytical problems set by intelligence tests may be much less effective at solving real-life problems that require creative thinking, or emotional maturity, or deep issues to do with the meaning and purpose of life. Thus we now find ourselves (rather like the nuclear physicist who has to propose new particles in order to explain results

that do not fit in with his theories) having to invent other forms
of intelligence such as 'emotional intelligence' (Goleman 1996)
and 'spiritual intelligence' (Zohar and Marshall 2000).

The Case for the Transpersonal

If all this sounds rather hard on science and on psychology, it is
not intended to be. Both science and psychology have been enormously successful at doing certain things, and the point is not
that they should be vilified but that we should allow ourselves to
recognise their limitations, and to acknowledge that some of the
criticisms they direct at religion also apply to themselves. If we
do so, then we are left with the realisation that there is a major
area of enquiry that lies outside their remit and that remains to
be addressed. This is the area into which religion falls, but it is an
area by no means confined to religion. The generally accepted
term for it is now the transpersonal. It is not an easy area to
define in a few words, but essentially it is that segment of human
thought and experience that transcends the narrowly personal. It
embraces spirituality and religious experience, mysticism,
altered states of consciousness such as Maslow's peak experiences (Maslow 1976), the paranormal in all its aspects including
Near Death Experiences (e.g. Ring 1985, Bailey and Yates 1996),
altruism in which the needs of others are placed above those of
self, unpossessive love, self-actualisation, meditation, contemplation and prayer, psycho-spiritual traditions of both East and
West, and the search for meaning and purpose in life together
with the practices of self-exploration and self-development that
accompany this search.

Investigations into the transpersonal represent arguably the
oldest branch of human enquiry, featuring as they do in the
teachings and writings of Pythagoras, Socrates, Plato, Aristotle,
Seneca, Epictetus, Marcus Aurelius, and many of the other leading figures credited with inspiring the vital stream of ideas on
existence and ethical action which enriched the civilisations of
ancient Greece and Rome, and helped shape the course of Western history. Transpersonal concerns were also central to the lives
of a large number of the men and women subsequently instrumental in shaping Western thought and belief — for example
Jesus of Nazareth, Origen, St. Paul, Dionysius the Areopagite, St.
Augustine, Boethius, Abelard, Eckhart, Aquinas, Luther and

Newman, Philo, Plotinus, Porphyry, Francis Bacon, Descartes, Spinoza, Berkeley, Kant, Leibniz, Hegel, Bergson and Whitehead among many others (Fontana 2004).

If we add to these two lists the names of noted thinkers from Eastern traditions also engaged in the transpersonal such as Lao Tzu, Buddha Sakyamuni, Mahavira, Shankara, Ramanuja, Nimbarka, Madhva, the Zen Patriarchs, Tsong Khapa, Nagarjuna, Ramakrishna, and Radhakrishna, we find that for more than 2,000 years transpersonal concerns have engaged some of the finest minds. The first signs that the scientific neglect of the transpersonal that characterised the Western world from the Age of Enlightenment to the second half of the twentieth century was coming to an end can be traced to the 1960s when Abraham Maslow and others began to develop transpersonal psychology (see e.g. Maslow 1968, Sutich 1969, and more recently Tart 1992, Walsh and Vaughan 1993) as a new movement that not only acknowledged the importance of the transpersonal (for Maslow it was clear that it represents man's 'higher and transcendent nature', and that this nature 'is part of his essence'), but that proposed it could be investigated in an academically rigorous way. They particularly emphasised the importance of recognising the inter-dependence of all living systems, and the need for an holistic approach to the investigation of these systems in contrast to the reductionist methods adopted by much of modern science. They also stressed the importance of introspection in the study of human psychology, reversing the suspicion with which it had been regarded by psychologists since the rise of behaviourism in the 1930s. Introspection, the ability of people to reflect upon and articulate their own experience, is an essential research tool if we are to investigate the inner life and much of what we consider makes us human, an issue to which we return at the end of the chapter.

From the first, transpersonal psychology has attracted outstanding psychologists and psychiatrists, many of whom have also distinguished themselves in more mainstream areas of enquiry. An abbreviated list includes Ken Wilber, Roger Walsh, Charles Tart, William Braud, Stan and Christina Grof, Frances Vaughan, Michael Murphy, Ian Gordon-Brown, James Hillman, Jack Kornfeld, Kenneth Ring, Daniel Goleman and Stephen LaBerge. Transpersonal approaches — again with an emphasis upon the inter-connectedness of all life and the dangers of studying phenomena in isolation from each other — have also been increasingly developed from the 1960s onwards in other scientific areas including physics, biology, medicine and engineering by leading figures such as Raynor Johnson, Sir John Eccles, David Bohm, Willis Harman, Jack Liggett, Fritjof Capra, John Mack, Brian Goodwin, Paul Davies, Edgar Mitchell, Michael Murphy, Glen Schaefer, Lary Dossey, Amit Goswami, James Lovelock, Rupert Sheldrake, Fred Wolf, Danah Zohar and many others.

However, the man who has arguably done most to integrate the various approaches within the transpersonal movement, and to identify the methodological differences between these approaches and those used in orthodox science and psychology is physicist and psychologist Ken Wilber (e.g. 1977, 1983, 2001a, 2001b). Wilber insists that our endeavours to reach an understanding of the universe and our place within it fall (and have always fallen) into two distinct methodological and conceptual groups, and it his identification of these groups and their respective strengths and limitations that offer us the best hope of formalising our thinking about the transpersonal on the one hand and orthodox science on the other, and retaining our respect for both forms of endeavour (Wilber 1998). The first of Wilber's two groups, Group One, has to do with the inner aspects of the individual and of the social collective and represents the transpersonal, while the second, Group Two, focuses upon the outer aspects of the two sets of phenomena and represents orthodox science. In the context of the individual, Group One is concerned with mind, intention, subjectivity, truthfulness and sincerity; and with culture, ethics, morals, worldviews, justice and understanding in the context of the collective. By contrast Group two focuses at the individual level upon observable behaviour, objective truth and its representation, and at the collective level upon society, objective nature and empirical forms generally.

Wilber emphasises that all the objects of study contained in Group Two are located in space — and therefore directly observable and quantifiable — while all those in Group One are not. The objects of study in Group Two are accessed through sensory perception, observation, measurement, science and the articulation of propositional truth, while those in Group One are accessed through interpretation, introspection, self-expression, art, aesthetics, common context, and inter-subjective meaning. Both Groups, Wilber insists, are of equal importance in any attempt to understand the totality of creation and to enhance human development. Each Group contains its own truths, and in each Group these truths can only be revealed by methods of investigation specific to itself. At no point is either Group reducible to the other.

The Two Groups at the Individual Level

In the context of the individual, the distinction between the two Groups and their methods is clearly evident if we look at the mind-brain controversy — the disagreement as to whether mind and brain are one and the same or distinct from each other. Arguably, no area of human study is of more crucial importance than this. If mind and brain are the same, then humans are no more than their biology, and no talk of many of the issues associated with Group One such as the spiritual and non-physical dimensions, religion and a possible afterlife — in short of the transpersonal — makes any sense. However, if mind and brain are not the same, then we are indeed justified in talking about and investigating such things. Unfortunately another popular myth obtains in this area, namely that through the use of Group Two methods psychology and neurophysiology have demonstrated that mind and brain are indeed one and the same. Modern research has shown that specific areas of the brain are associated with specific mental events, and this is assumed by many scientists and psychologists to mean that as the brain is involved in these events, it therefore must be the originator of them.

This assumption emerges clearly from the dialogues arranged between top Western scientists and the Dalai Lama (Houshmand Livingstone and Wallace 1999, Wallace 2003a). There is no readiness on the part of the scientists involved in these dialogues to recognise that if our knowledge of the brain-mind relationship is based simply upon correlations between brain activity and mental events, then the question whether brain generates mind or whether mind is distinct from brain and simply works through it has something of a chicken and egg quality about it. If brain generates mind then certainly we would expect specific areas of the brain to appear activated when mental events take place, but we would equally expect these areas to be activated if mind works through brain. Irrespective of whether brain creates mind or whether mind works through brain, the brain is the agent through which mental events interact with the physical world.

The argument that we do not know how a non-physical mind can interact with a physical brain (clearly a major stumbling block for many scientists) does not invalidate the fact that it may do so — because we equally do not know how the electro-chemical activity of the brain can produce non-physical events such as thoughts, intentions, volitions and a moral sense. In addition, despite our ignorance as to how a non-physical mind can interact with a physical brain, we do have tentative hints from two different sources that mind nevertheless does appear able directly to influence physical matter. Both sources give rise to issues that are far too complex to be more than touched on here, but the first of them is quantum mechanics where findings have emerged that suggest that consciousness, through the simple act of observation, may affect the behaviour of sub-atomic particles (see e.g. Goswami 1993, Gribbin 1998). Such findings remain controversial, not least because we do not have adequate language or concepts for dealing with the mysteries of the quantum world. Nevertheless, it does seem possible that at the level of the quantum vacuum (which exists inside every atom, including those that make up our bodies and our brains) interaction between a non-physical mind and a physical brain may be a possibility. Alternatively, we may perhaps suppose that at this level the distinction between mind and matter disappears altogether, and we are left with a universal energy field, a field inseparable from the forces that brought the cosmos into being.

The second source is psychical research, which over the last century has generated evidence that supports the existence of extra sensory abilities that so far defy explanation in terms of the known laws of science (e.g. Radin 1997). Let me take just one example from this extensive body of knowledge, namely that described recently by Braud (1997). Braud, one of the leading figures and most careful scientists involved in psychical research, details research conducted by his own team which indicates that some individuals are able to increase or decrease the autonomic nervous activity of other people mentally and at a distance. An impressive review of the findings allows him to conclude that 'our states of mind and conditions of being' can indeed have important and direct influences upon the actions and being of others. Since the people influenced in this way were not conscious of this influence, and since the autonomic nervous system is inextricably linked to the material body, the inference is that here again we may have evidence of a non-physical mind directly influencing physical matter.

On the basis of Braud's findings we can suggest that our own minds may influence our own bodies (including the brain) by whatever non-physical means allows some minds to influence the bodies of other individuals. Braud refers to this non-physical means as 'spirit', and in his closing paragraph he affirms that:

> Spirit permeates the material and impacts upon all facets of our lives, both exceptional and mundane — upon our bodies, our emotions, our relationships, and our expressions of creativity.

This is nothing short of a paradigm-shaking statement, and it is something of a commentary upon modern Western culture that such a statement, made by a highly respected scientist and on the basis of meticulous research findings, fails to make headlines when the media gives so much space to so much less worthy issues.

However, irrespective of the insights provided by psychical research and tentatively by quantum mechanics, it is correct to insist that in terms of Wilber's two Groups, mind belongs to the first and brain belongs to the second. Mind belongs to the inner world and is experienced only by its owner, brain belongs to the outer world and can be observed objectively by others. No one interested in truth can afford to overlook this fact. Mind is studied by the reports people give of their own inner world of thought, dreams, concepts, valuations etc., while brain is studied by the outer exploration of electro-chemical activity in the cerebral cortex. Thus the operational if not the actual distinction between mind and brain is self-evident, as is the fact that the two require to be investigated by quite different methodologies.

The Two Groups at the Collective Level

The distinction between the two Groups at the collective level should also become clear from a simple example. If we wish to know the reasons why people attend church, the inner approach identified with Group One would be to ask them why they do — to which they might answer variously that they do so in order to worship God, to pray for those they love, to assure themselves of an after-life, and to experience what they might describe as spiritual uplift. The outer approach identified with Group Two would be to compare the behaviour of church-goers with that of non-churchgoers — and to discover for example that church-goers enjoy more social support, show less reliance upon alcohol and other stimulants, have a lower incidence of physical and mental illness, and commit fewer civil and criminal offences.

Both Group One and Group Two approaches employ valid methodologies and produce valuable data. Both sets of methodologies have their place in any investigation. But neither set can be substituted for the other. Our relative neglect of Group One in favour of Group Two at both the individual and the collective level has led to an imbalance not only within science itself but within Western culture in general, the result of which has been that although we may know more and more about the outer world, we know less and less about ourselves. We spend a great deal of time exploring the outer world revealed to us by our minds, and little or no time in exploring those minds themselves. Our orientation as a culture has become increasingly superficial as we are taught through the educational process and the media to look outwards rather than inwards for our discoveries, for the solutions to our problems, for our entertainment and amusement, and for answers to the fundamental questions about life and its meaning.

It was not always thus. Prior to the eighteenth-century Enlightenment and the Age of Reason, the balance between the two groups was tipped in the other direction. Attempts to study the outer world and to develop perceptual science were seen as heresy by the Church, which regarded all such knowledge as provided by revelations that came exclusively through the inner world. Thus an imbalance between the two groups, no matter which of them is favoured, appears damaging, not least because it allows those who control the dominant group to behave with intolerance towards the other group, and with an arrogant insistence that it possesses a monopoly of the truth about both inner and outer worlds.

Exploring the Inner World

As made clear earlier, introspection is regarded within transpersonal psychology as an essential tool in exploring the mental phenomena associated with Group One. Whereas Group Two neuroscience, which studies the brain as a passive organ that merely processes information fed in from the outside, has no explanation for how a sense of unified 'self' can arise out of the many trillion parallel processes taking place in the brain at any one time, introspection demonstrates the existence of this 'self' through direct experience. Furthermore, through the various techniques of mind training such as meditation developed in both Eastern and Western traditions, introspection can enable the individual to explore deeper and deeper into the essential nature of this 'self' in all its many aspects. As Wilber points out, by identifying similarities between the experiences reported by those carrying out this introspection, we can reach perfectly acceptable scientific conclusions about the way in which the mind operates at its many different levels. As already discussed, one of the basic tenets of the transpersonal approach is the holistic nature of reality. Thus, when studying the deeper reaches of the mind through introspective methods, we would expect to find impressive levels of consensus among the experiences reported, and this indeed has proved to be the case. Eastern psycho-spiritual traditions such as Hinduism and Buddhism have in fact built up complex and impressive models of the mind through the use of introspection that are far superior to anything as yet achieved by Western science (e.g. deCharms 2000).

Conclusion

I must stress again that I am in no sense dismissing the disciplines identified with Group Two. Western science and Western psychology have highly impressive achievements to their name. Nor am I seeking to favour the disciplines associated with Group One at the expense of those in Group Two, or to suggest that Group One methods can be used to answer Group Two questions. My emphasis is upon the fact that many of the claims and some of the philosophies associated with Group Two are misplaced and at times dangerously misleading. Equally dangerous and misleading is the prevailing rejection of Group One disciplines as unscientific and philosophically suspect. There is a grave error in supposing that the methods of modern science are the only way of answering questions about existence, or that in many areas they are necessarily the most appropriate way.

The current imbalance between Group One and Group Two promises to be redressed in some part by the development of transpersonal psychology and other transpersonal disciplines, using methods of enquiry as rigorous in many ways as those employed in Group Two. This is not simply an academic matter. A recognition of the inter-connectedness of all things, emphasised by the transpersonal movement, is vital not only to our understanding of the nature of existence but to our very survival as a species. The separation that we have created not just between the self and other but between humanity and creation lies at the heart of the desperate situation to which we have reduced not only ourselves, with our incessant warfare and crimes against each other, but our world and all it contains. If we persist in this philosophy of separation, and in the concomitant inability to recognise that we destroy ourselves as we destroy other species and the very environment on which we depend, then the demise of the human race would seem inevitable, just as we have already brought about the demise of so many species who intended us no harm. When this destruction occurs — as assuredly it will unless we mend our ways — far from being able to project our guilt outwards in the modern practice of looking always elsewhere than within ourselves, we will have no option other than to recognise that alone of all species it is we humans who have pulled the roof of the temple down upon us all. That is assuming that any of us survive long enough to acknowledge our guilt.

References

Bailey, L. W. and Yates, J., ed. (1996), *The Near-Death Experience*. New York and London: Routledge.

Braud, W. (1997), 'Parapsychology and spirituality: Implications and intimations', In C. Tart (ed.) *Body, Mind, Spirit: Exploring the Parapsychology of Spirituality*. Charlottesville, VA.: Hampton Roads.

Boyer, T. (1985), 'The classical vacuum', *Scientific American*, August.

deCharms, C. (1997), *Two Views of Mind: Abhidharma and Brain Science*. New York: Snow Lion Publications.

Fontana, D. (2003), *Psychology, Religion and Spirituality*. London: Blackwell.

Fontana, D. (2004), 'The development and meaning of transpersonal psychology', *Transpersonal Psychology Review* (in press).

Goleman, D. (1996), *Emotional Intelligence*. New York: Bantam.

Goswami, A. (1993), *The Self-Aware Universe*. London and New York: Simon & Schuster.

Gribbin, J. (1998), *Q is for Quantum*. London: Weidenfeld & Nicolson.

Houshmand, Z., Livingstone, R.B., Wallace, A.B., ed. (1999), *Consciousness at the Crossroad: Conversation with the Dalai Lama on Brain Science and Buddhism*. Ithaca, NY: Snow Lion Publications.

Maslow, A. (1968), *Towards a Psychology of Being*. Princeton: Van Nostrand (2nd edn.).

Maslow, A. (1976), *The Farther Reaches of Human Nature*. Harmondsworth: Penguin.

Radin, D. (1997), *The Conscious Universe: The Scientific Truth of Psychic Phenomena*. New York: HarperCollins.

Ring, K. (1985), *Heading Towards Omega: In Search of the Meaning of the Near-Death Experience*. New York: William Morrow.

Sutich, A.J. (1969), 'Some considerations regarding transpersonal psychology', *The Journal of Transpersonal Psychology*, **1** (1), pp. 11–20.

Tart, C., ed. (1992), *Transpersonal Psychologies: Perspectives on the Mind from Seven Great Spiritual Traditions*. San Francisco: Harper.

Wallace, B.A. (2003a), *Buddhism and Science: Breaking New Ground*. New York: Columbia University Press.

Wallace, B.A. (2003b), *Choosing Reality*. Ithaca, NY: Snow Lion.

Walsh, R. and Vaughan, F., ed. (1993), *Paths Beyond Ego: The Transpersonal Vision*. Los Angeles: Tarcher.

Wilber, K. (1977), *The Spectrum of Consciousness*. Wheaton, IL: Quest Books.

Wilber, K. (1996), *Up From Eden: A Transpersonal View of Human Evolution*. London: Routledge & Kegan Paul.

Wilber, K. (1998), *The Marriage of Sense and Soul: Integrating Science and Religion*. Dublin: Newleaf.

Wilber, K. (2001a), *A Brief History of Everything*. Dublin: Newleaf.

Wilber, K. (2001b), *A Theory of Everything*. Dublin: Newleaf.

Zohar, D. and Marshall, I. (2000), *SQ – Spiritual Intelligence*. London: Bloomsbury.

Keith Ward

Human Nature & the Soul

Introduction —
Traditions about the Soul

Traditional views of human nature have been concerned with the idea of the soul, and with making an important distinction between humans and other animals. There is, however, a very great difference between official doctrines of the human soul and popular understandings of it. I will try to bring out that difference, and then ask whether, and in what sense, there remains an important distinction between humans and other animals which some doctrine of the soul might help to mark.

There are two major philosophical traditions about the soul, one Indian, based on the Vedas and Upanishads, the other Semitic, based on the Bible. There are, of course, many other traditions, especially in China and in many societies based on oral traditions, but it is the Semitic tradition, shared between Judaism, Christianity and Islam, with which I shall be concerned here. I shall refer to the Indian tradition only briefly and by contrast with the Semitic. It is worth making such a reference, however, because views related to the Indian tradition are quite widespread in Europe and North America, and tend to get elided with the Biblical tradition, in a rather confusing way.

It is usual in India to distinguish spirit (*purusa*) and body or matter (*prakriti*). The basis of all reality is spirit, usually said to consist of being, consciousness and bliss. Each human person is a primarily spiritual reality, which exists without beginning or end, and can be embodied in many different forms of matter, or bodies. Thus I, as an individual self, have never begun to exist, nor will I ever cease to exist (cf. *Bhagavad Gita*, chapter 2, verses 11–13). But this particular personality, which is associated with the body I now have, will cease, and I will have many other bodies, probably a countless number, since in endless time I will have many embodiments. The general aim of much Indian spirituality is to achieve liberation from this round of rebirth, and realise one's true nature as pure spirit.

I mention this because many people seem to think that 'the soul' is that part of a human person, the spiritual part, which can have many different bodies, which continues after death and which existed before birth, and which can be disconnected from the body without any harm to the personality. The soul, in other words, is a spiritual substance which is better off without a body, and the body is like a suit of clothes which you can put on and take off again, without changing underneath.

Aristotle — The Soul as a Formative Principle

In some of Plato's Dialogues, this view can be found, and because one of the most influential early Christian writers, Augustine, was heavily influenced by Plato, it has sometimes been popularly thought that Christians also think that the soul is a spiritual substance, quite distinct from the body. However, this would be a major misunderstanding of Augustine, and also of Descartes, whose view was basically identical with that of Augustine. The Christian tradition to which they both belonged, like the Jewish and the Muslim, starts at quite a different place.

That place is clearly set out in the early chapters of the Book of Genesis, where it is said that 'the Lord God formed man of dust from the ground, and breathed into his nostrils the breath of life' (Genesis 2.7). Human persons are formed of dust, into which breath (*neshamah* or *nephesh*) is breathed by God. Whatever the soul is, in the Semitic tradition, it is not a substantial entity which exists without and before the body, and is better off without a body. In fact, although the Hebrew word *nephesh* is often used for soul, it might be better not to use the word soul at all and just to speak of the breath of life. This would make clear that the soul, for Hebrew thought, is the active principle of a living body. It is not a separate entity, additional to the body.

Now just what this 'breath of life' is has always been open to discussion. In the development of Jewish and Christian thought, it was Aristotle, and not Plato, whose philosophy was used to build up a doctrine of the human soul. Aristotle, in his work *Peri Psyche*, 'On the Soul', called a soul the Form of a living body. This idea of 'Form' (*eidos*) has itself become alien to modern thought. Aristotle divided all material things into matter and form. Roughly, the matter is the stuff of which things are made, and the form is the defining property which makes a thing what it is and not another thing. For Aristotle, a Form is a real constitutent of things, a dynamic principle which gives particular existence to matter, a sort of blueprint for building chunks of matter in particular ways. But it is an active blueprint, with a causal input into the nature of things. So we might think of an Aristotelian Form as a dynamic principle which shapes matter to a specific pattern. It is not another chunk of matter, not even very thin matter, ghostly substance. But it is a real causal principle that enables pieces of matter to have a character or nature.

Every particular thing has a Form, even a rock or a star. But some Forms are more complex, they cause more integrated patterns of organisation, and include simpler Forms in wider and more inclusive organic wholes. These more inclusive integrated patterns of organic structures are called 'souls' by Aristotle. There are vegetable souls, the dynamic causal principles which shape matter into rich and enduring organic wholes. They build up their structures by using energy from the environment, and they reproduce those structures. Then there are animal souls, which add another layer of complexity to the process, enabling animals to respond to their environment by sense-awareness, and to initiate movement to increase the opportunities for growth and reproduction. And there are what he calls intellectual souls, which are the highest organising principles known to us on the planet. Intellectual souls generate the capacity in matter to imagine and reflect on information obtained from the environment, and to plan courses of action leading to envisaged goals which are not restricted to the effects of immediate stimuli. Mathematical thought and responsible action are the marks of the intellectual soul, the highest, most inclusive, integrating principle of matter known to us.

Now all this may seem to be out-dated science. Its interest is mainly that it provided the conceptual basis for the traditional European idea of the soul, when it was developed in Jewish and Christian thought forms. What should be quite clear is that Aristotle is in some subtle sense a materialist, or perhaps better a mind-body monist, or what is sometimes called a double-aspect theorist. He does not think that matter can exist without Forms, and he does not think that Forms can exist without matter. Both Form and matter are necessary to the existence of any actual thing. So there is no question of the soul existing without a body. The soul is the shaping and organising principle of a certain sort of living body, a complex, organised material structure. It comes into existence with the body, and presumably dies with it, too.

It should also be clear that the distinction between humans and other animals is not that humans have souls and other animals do not. There is a distinction between sorts of souls, which is largely one of degree — the degree of complexity, consciousness and possibility of free and responsible action. Certainly, intellectual souls (we might say, beings with the capacity to think abstractly and decide freely) have different capacities than souls which simply respond to stimuli or do little more than move and reproduce. There are lines to be drawn between types of activity and agent. But these lines may be very fuzzy, with many border cases, and they might or might not be drawn between members of distinct biological species. For instance, the question of whether the great apes have intellectual souls is a matter for empirical investigation of their capacities, not of *a priori* definition.

Aristotle on Human Nature

The Aristotelian answer to the question, 'What is human nature?' would be that human beings are chunks of matter which possess distinctive capacities, specifically those of abstract thought and responsibly free action. The proper virtue (or excellence) of a human being is thus to realise the activity of abstract thinking (*theoria*) and morally responsible action, and to do so in a community of similar rational agents, since humans are essentially social animals which live in community with one another.

Who can be a member of this human community? Anyone who can enter into the social processes of intellectual reflection and moral decision-making — which so far limits membership to mature humans (not in principle, but merely in fact: there could be other members, non-human persons, but there do not seem to be). But we would have to include, too, all who are in principle capable of such social reflection and action, but who do not actually realise their capacities because of youth, handicap, disease, or age. Because Aristotle thinks in terms of capacities, and not just of actual achievements, any being will be deemed to have a human nature who has the Form, the capacities of an intellectual being, even if they are not able to be exercised for various reasons.

At this point an Aristotelian view would differ radically from any view that a person can only be a being who is actually intelligent and responsible. For Aristotle, a human mentally disadvantaged baby has a different nature from a highly intelligent chimpanzee, a different patterning principle laying down distinctive capacities for a piece of matter, and so should be treated as an intellectual soul, not as an animal soul. Of course, this leaves open the question of how intellectual and animal souls should be treated. It should be remembered that Aristotle was able to justify slavery and the radical inferiority of women, so there can obviously be disputes about how intellectual souls should be treated, and about whether there are relevant distinctions within the class of intellectual souls. But the general principle remains that how you treat things will depend on the natures they have — you should treat a thing in a way appropriate to its nature, not just to its individual attainments, and you discover the nature by observing mature normal cases of that kind of thing.

There are, then, three important basic points for an Aristotelian view of human nature. First, humans are fully and properly material objects. Second, their distinctiveness consists, not merely in that they are members of the human species, but in their having distinctive mental capacities of thinking and acting. Third, all human organisms should be considered as having human nature, and so should be treated in accordance with possession of such a nature, not just in accordance with their individual achievements (there is a place for treatment in accordance with achievements, but that lies above the basic moral level of 'human rights', and complicates things in a way I do not wish to get into at this point).

Aquinas and Descartes:
Soul as 'Substantial Form'

When this view was taken into Christianity — and it could be said to have been officially authorised in the work of the thirteenth century philosopher Thomas Aquinas — some amendments had to be made, but they are not radical ones. In the first place, a Christian might well be unhappy to be a materialist. Since Christians think there is a non-material creator of the universe, there must be the possibility of a non-material existent. Further, this existent will have some personal characteristics — at the least, consciousness and intention, if the universe is generated by something which, however remotely, can properly be called an act of will. So the non-material, or spiritual, will ultimately have causal priority over the material. So Christians have to make it clear that an immaterial subject of consciousness and will is possible.

It might be pointed out that Aristotle did not exclude this possibility, and in the twelfth book of the Metaphysics, outlines what has become the classical Christian idea of God as the supremely desirable being resting in blissful awareness of its own perfection, and drawing all things into likeness with itself simply by its own supreme desirability. It is particular things within the universe, compounded of matter and Form, which have a material nature, and this is compatible with their having a purely spiritual cause.

Augustine and subsequent Christian thinkers developed Aristotle's thought to say that God not only drew the universe towards the divine perfection by attraction, or, as Aristotle said, by being loved, but God actually created the universe, that is, actively willed it to exist. Moreover, in the incarnation God entered into the universe, and united human nature to the divine nature, so that God could raise that human nature to participate in some way in the divine perfection.

The consequence of these ideas is that it is slightly odd to call a Christian view 'materialist', even though human beings certainly are lumps of highly organised matter. What is important about human beings, for Christians, is that as well as being material things which can reflect and decide responsibly, they can come to know and love a purely spiritual reality, God. Further, consciousness and will are, in their primary instances (i.e. in

God), immaterial realities. So one might expect that consciousness and will, even where they exist in finite beings, will at least
have an immaterial aspect, perhaps that they will be primarily
psychic elements or realities. In most cases, those psychic elements may be dependent upon the physical beings of which they
are non-physical parts. But at a certain stage of development
(and certainly in humans) they come to have the capacity to
interact with a purely immaterial being.

So in Aquinas one finds that the intellectual soul, though it is
the principle that gives to matter the capacities of a thinking
being, has a sort of substantial, though adjectival, existence. It is,
he says, a 'substantial Form' (*Summa Theologiae*, 1, question 76.
article 1). This is, incidentally, the view of the much-maligned
Descartes, whose extreme sounding dualism is mainly part of a
thought-experiment of 'hyperbolic doubt', which he himself
overcomes by accepting the common-sense view that the soul is,
as he puts it, 'intimately united with it (the body), and so confused and intermingled with it that I and my body compose, as it
were, a single whole' (*Meditation 6*).

What Aquinas means is that each soul is the integrating principle which gives mental capacities to a particular body. It originates with the body. To be precise, Aquinas thought, following
Aristotle, that the intellectual soul originates 60 days after conception. Before that time human embryos have vegetable or animal souls, and at 60 days God destroys those and replaces them
instantaneously with a newly created and completely individual
intellectual soul. Each soul is particularised to shape one specific
body, and it will be incomplete without the body, which it is
meant to shape.

Muslim, Jewish and Catholic
Views of the Soul

The Muslim medieval view, which has not subsequently been
amended, is that the intellectual soul is 'breathed into' the body
at 120 days. The general Jewish view is that properly human life
begins when the head of a baby emerges from the birth canal.
Neither of these views should be taken to imply that abortion is
permitted before that time, though it is relevant that a human
soul is not taken to exist before then.

The official Roman Catholic view at present is that since no-one knows when the intellectual soul is created, one ought to play safe, and act as though properly human life may begin at conception. But there is no dogmatic statement about when the soul is created. Other Christians are free to make their own decisions about this, but it is of some interest that the religious traditions were originally prepared to place the onset of human nature quite late in embryonic development (for Aquinas, when the embryo had a recognisably human form).

Clearly there is something substance-like about this soul, since it comes into existence at a specific time — and, according to Catholic doctrine, is freshly and immediately created by God. But it is still a principle for shaping a body; that is its proper job. It is emphatically not the true essence of a person, which can exist fully and properly without any body at all, or with a number of different bodies.

Nevertheless, since the activity of the intellectual soul issues in consciousness and will, it issues in states of affairs which are not necessarily material (even though it is some body which has consciousness and wills things to happen). So in the thought of Aquinas there is an immaterial principle of organisation, and there are immaterial states of consciousness and will. A logical space opens up (and Aristotle had noticed this) in which the principle could operate over immaterial states alone.

To put it crudely, if this body generates thoughts and memories, could the capacity for thinking and deciding, and the subject in which those capacities are grounded, not operate simply on those contents of consciousness, without a material body? Aquinas says that it could do so, but would do so 'improperly and unnaturally'. We might have attenuated persons, psychic structures of thought and memory existing in a disembodied state. But, as Aquinas says, 'my soul is not me'. That is, if my conscious contents continued to exist after the death of my body, and the principle of thought and action that had actuated me continued to operate on them, something would exist which had a very close connection with me, but it would not really be the person that I am. If this person is to exist after bodily death, then my body must in some way be resurrected, reassembled, or replicated, so that I could be a whole person again. The resurrection of the body, for the Semitic view, is essential if there is to be personal immortality.

Resurrection of the Body

I think that Christians are committed to the possibility of the human person continuing to exist after the death of the body. When Paul talks about the 'resurrection body' in 1 Corinthians 15, he does not talk about disembodied souls, and he does not talk about ordinary physical bodies brought back from the dead. He talks of what he calls 'spiritual bodies', which are incorruptible, and whose form is quite different from the flesh and blood we now have which, he plainly says, 'cannot inherit the kingdom'. Wherever such embodiment takes place, it cannot be within this space-time, which is inherently corruptible. We might put this by saying that this body generates, from its central nervous system and brain, many conscious states, many memories, intentions, and states of knowledge. It generates a continuing self, which accumulates experiences and develops projects over long periods of time. That self is a subject of experiences and actions, and to be such it must be embedded in an environment from which it receives information and over which it has some control, usually through one localised part of the environment. This is its body. Without a body, there will be no way we can imagine of gaining new experiences and undertaking new actions. So selves are naturally embodied. But perhaps memories, knowledge and learned capacities can be transferred to another form of embodiment.

The idea of resurrection is the idea of a new form of embodiment which is so closely related to this one that it can be best spoken of as the 'same body transformed', rather than as quite a new body. We may think of it as some form which enables our unique personal capacities to be more creatively and freely expressed than in this corruptible world. But, as Paul reminds us, what such bodies will actually be like is unimaginable by us now.

One reason this doctrine is popularly misunderstood lies in the promulgation of three Catholic doctrines, that the soul is directly created by God at a specific point (which makes one think it is a thing), that the soul is naturally immortal (which makes one think I could just go on for ever without a body), and that the body is resurrected (which makes one think that the very same body climbs out of the grave). To get the orthodox view, one has to have some grasp of the idea that the soul is the Form of the body, and yet that it is a substantial Form, capable of (un-naturally) existing with just conscious contents. When God creates a soul, God creates a subject with capacities and dispositions which are properly expressed in and through a particular material body. When God gives a soul immortality, God gives it the un-natural possibility of existing in an attenuated sense between the death of its body and the resurrection of that body. And when God resurrects a body, God creates a fully appropriate vehicle of expressing the acts and experiences of the same subject that were expressed in a partly impeded way in the material body in this space-time environment. On this view, the soul is clearly not a complete independent entity, a 'substance'.

If one gives up the philosophy of Aristotle, this view will have to be reformulated. The idea of a Form as an actual integrating principle looks suspiciously like biological vitalism, and though there may be a surprising amount of life left in it, it does seem to go against the grain of most recent biological thinking. But perhaps one can drop the philosophical idea of Form, and retain what is central to Christian thinking on this issue. What seems central to the tradition is that talk of the soul is talk of the capacities of an agent and experiencer, which is properly embodied and expressed in a material substance, which both limits and makes possible the expression of its inherent capacities. This agent, the personal subject of consciousness and intention, can continue to exist in a secondary sense after the death of the material body. But any worthwhile and properly personal immortality will involve at least some analogue of material embodiment.

If we wanted to make an attempt to restate a broadly Biblical view in a post-Aristotelian way, I consider that the best way to do this would be to speak of the evolutionary process as one in which emergent properties of consciousness and intention develop from earlier propensities to register stimuli and respond to them, in simpler forms of organism. We might not want to draw a clear line between human, conscious beings, and non-human, unconscious beings (this was a mistake made by Descartes, and unfortunately not absent from all Christian thinking). We might rather say that there are many degrees of emergence, from unconscious rocks through the sense-awareness of lower animals to the capacities for abstract thought and self-legislated, rule-governed action which are primarily found on earth in communities of human beings.

The Genesis account does not suggest a total difference of kind between human and other animals. All have the breath of life (Genesis 1.30), and it is not unique to humans to have either organic or mental properties. What Genesis does is to assign a unique role to humans: they are created in the image and like-ness of God (Genesis 1. 26). The usual early Christian interpreta-tion of 'being created in God's image' was to say that humans have reason and freedom, as God has. More recent Hebrew scholarship rather stresses the human capacity for relationship, especially but not exclusively relationship with God. Perhaps classical Christian thought focussed too much on the intellectual capacities of humans, and not enough on the emotional and rela-tional. The sensitivity, creativity and co-operative capacities of humans are as important as their ability to do pure mathematics or science. But in any case what distinguishes human beings is that they are able to exercise a rational and responsible control of and care for the natural world of which they are an integral part.

The human role is to care for the earth, to order it as God ordered creation out of chaos, so that 'being in God's image' is a matter of having the role of caring for creation as God does, and assuming responsibility for its ordered flourishing. The claim (popularised and partly accepted by historian Lynn White) that Genesis licenses the exploitation of nature is the opposite of the truth, which is that it commands the care of nature. Humans are to have dominion, in the sense of Lordship under God — and as God created nature, that means to care for nature on behalf of God.

The Biblical account thus suggests that human distinctiveness lies more in responsibility for the natural world than in exemp-tion from its dictates. From a theological point of view, having a human nature consists in the ability of humans consciously to cultivate and develop the likeness of God in themselves. It is not a matter of uniquely being spiritual substances in the middle of a mechanistic, soulless universe. It is a matter of being able con-sciously to realise capacities which have been potential in the material cosmos from its beginning, and which have emerged gradually through ever more complex forms of organisation within the universe.

In what sense does this mark out human life as having special dignity among other creatures of God? It is certainly not that only human beings are worthy of respect. It should always have been obvious, though unfortunately it has not, that if God creates the universe, it must be respected precisely as something chosen by, and therefore presumably valued by, God. The chief reason for a Christian to value nature is that God values it, and because it is created by God it has intrinsic value of its own.

Human beings are distinctive in having the capacity to be conscious of cumulative experience and to have intelligently directed feelings and actions. Throughout the course of their lives they build up a store of experience which includes understanding and sympathy. They are able to realise purposes and goals over time, to shape their own characters in new ways. They are continuing subjects of complex cumulative experience and time-spanning action. It is perhaps only when material beings have such a cumulative character of experience and a capacity to be responsibly self-directing that it becomes meaningful to ascribe to them the possibility of immortality, a continuing sense of self beyond the existence of a particular body, and a conscious and developing relationship to God, who is a purely immaterial being.

Humans may have special dignity in being called to immortality, in being held accountable for their response to that call, and in requiring from others a respect for their freedom in making such a response. I believe that it is meaningful to speak of a distinctive human nature, and that it is morally important to do so. But it is equally important to see that the capacities and responsibilities which ground that distinctiveness are not essentially confined to the human species, and that humans are integral parts of a material world within which their role is to create and conserve the values which the creator has made implicit in the cosmos, to be stewards of a divinely valued creation, and not manipulators of a cosmic machine. Within the Semitic traditions, these are the values that speaking of 'the soul' should seek to safeguard, and these are the states and capacities which talk of the soul primarily denotes.

Mary Midgley

Mind & Body:
The End of Apartheid

Separate Substances?

What does it mean to say that we have got a mind–body problem? Do we need to think of the relation between our inner and outer lives as business transacted between two separate items in this way, rather than between aspects of a whole person?

Dualist talk assumes that we already have before us two separate things which we don't see how to connect. This is a seventeenth century way of seeing the problem. It is tied to views in physics and many other topics that we no longer hold.

'Mind' and 'matter', conceived as separate in this way, are extreme abstractions. These are terms that were deliberately designed by thinkers like Descartes to be mutually exclusive and incompatible, which is why they are so hard to bring together now. In Descartes' time, their separation was intended as quarantine to separate the new, burgeoning science of physics from views on other matters with which it might clash. It was also part of a much older, more general attempt to separate Reason from Feeling and establish Reason as the dominant partner, Feeling being essentially just part of the body. That is why, during the Enlightenment, the word 'soul' has been gradually replaced by 'mind', and the word 'mind' has been narrowed from its ordinary use ('I've a good mind to do it') to a strictly cognitive meaning.

That was the background against which philosophers designed the separation of soul and body. And they saw it as an answer to a vast metaphysical question of a kind which we would surely now consider ill-framed. This was still the question that the pre-Socratic thinkers had originally asked; 'What basic stuff is the whole world made of?' And the dualist answer was that there was not just one such stuff but two — mind and body.

In the seventeenth century, hugely ambitious questions like this were much in favour. Perhaps because of the appalling political confusions of that age, its thinkers were peculiarly determined to impose order by finding simple, final answers to vast questions through pure logic, before examining the complexity of the facts. In philosophy, as in politics, they liked rulings to be absolute. The grand rationalist structures that they built — including this one — supplied essential elements of our tradition. But there are limits to their usefulness. We do not have to start our enquiries from this remote distance. When we find the rationalist approach unhelpful we can go away and try something else.

How Consciousness Became a Problem

Officially, we English-speaking philosophers are supposed to have done this already over mind–body questions. Half a century back we agreed that we should stop talking in terms of a Ghost in a Machine. But our whole culture was much more deeply committed to that way of thinking than we realised. Existing habits made it seem that our next move would be quite simple. We could at last triumphantly answer that ancient, pre-Socratic question — which was still seen as a necessary one — by once more finding a single solution for it. We could rule that everything was really matter. We could keep the material machine and get rid of the mental ghost.

So behaviourist psychologists tried this. They tabooed all talk of the inner life, with the effect that, through much of the twentieth century, people who wanted to seem scientific were forbidden to mention consciousness or subjectivity at all. But this turned out not to work very well. A world of machines without users or designers — a world of objects without subjects — could not be made convincing. It gradually became clear that the concept of the Machine had been engineered in the first place to fit its Ghost and could not really function without it. Attempts to use it on its own turned out so artificial and unreal that the learned population eventually rebelled. Some thirty years back, scientists suddenly rediscovered consciousness and decided that it constitutes a crucial Problem. But the concepts that we now have for dealing with it are still the ones that were devised to make it unspeakable in the first place.

Colin McGinn has stated this difficulty with admirable force in his book *The Mysterious Flame* (1999):

> The problem is how any collection of cells ... could generate a conscious being. The problem is in the raw materials. It looks as if, with consciousness, a new kind of reality has been injected into the universe.... How can mere matter generate consciousness? ... If the brain is spatial, being a hunk of matter in space, how on earth could the mind arise from the brain? ... This seems like a miracle, a rupture in the natural order (pp. 13–15).

McGinn's drastic answer is that this state of affairs is indeed a real mystery — a puzzle that our minds simply cannot fathom because it lies outside the area that they are adapted to deal with. His suggestion is that there must be an unknown physical property, which he calls C*, that makes consciousness possible. This property is present in the stuff of brains, but it may be something which it is altogether beyond us to understand.

It is surely good news to find a respected analytic philosopher recognising mysteries — insisting that there are limits to our power of understanding. But I shall suggest that we don't need to fall back on his rather desperate solution. This particular difficulty arises from a more ordinary source. Our tradition is leading us to state the problem wrongly. We really do have to start again somewhere else.

I will suggest that a better starting-point might be to consider directly the relation between our inner and outer lives — between our subjective experience and the world that we know exists around us — in our experience as a whole, rather than trying to add consciousness as an afterthought to a physical world conceived on principles that don't leave room for it. The unit should not be an abstracted body or brain but the whole living person. In order to show why this is necessary it will be best to glance back first at the tradition to see just how and where things have gone wrong.

Rationalist Wars

This takes us back to Descartes. But of course he is not personally to blame for our troubles. If he had never written, sooner or later someone else would certainly have made the dualist move. And it is most unlikely that they would have done it better than he did.

As I have suggested, one factor calling for dualism was the general, lasting wish to establish Reason as a supreme ruler, a separate force able to arbitrate the confusion caused by disputes between warring authorities in the world. But the special factor that made this need pressing at that time was the advent of a new form of Reason — one that seemed likely to compete with old forms of knowledge — namely, modern physics.

Once that discipline was launched into an intellectual world that had been shaped entirely round theology — a world, too, in which theological opinions were dangerously linked to international politics — some device for separating these spheres had to be invented. That device ought to have been one that led on to Pluralism — meaning, of course, not a belief that there are many basic stuffs but a recognition that there are many different legitimate ways of thinking. Different conceptual schemes can quite properly be used to trace the different patterns in the world without conflicting. But, instead, the train of thought stopped at the first station — dualism — leaving its passengers still stranded there today.

We see signs of this trouble whenever people raise this kind of question — for instance over the problem of Personal Identity. When we talk about relations between mind and body, we are asking what a person essentially is. Modern analytic philosophers have puzzled a great deal about this, usually setting out from John Locke's discussion of it and concentrating on just one point in that discussion — his famous example of the Prince and the Cobbler (*Essay Concerning Human Understanding*, Bk. 2, Ch.27, Section 15).

Locke argued that, if we ask whether someone is 'the same person' as he was in the past, the answer must depend on the continuity of his memory, not on continuity of substance, 'For', (says Locke) 'should the soul of a prince, carrying with it the consciousness of the prince's past life, enter and inform the body of a cobbler ... everyone sees he would be the same person with the prince, accountable only for the prince's actions'.

Starting from this little example philosophers have produced a striking monoculture of science-fiction stories. They have repeatedly asked whether various kinds of extraordinary beings would count as 'the same person' after they had undergone equally extraordinary kinds of metamorphosis. Their answers tend not to be very helpful because, when we go beyond a certain distance from normal life, we really don't have a context that might make sense of the question. And — as students often complain — these speculations seem fairly remote from the kind of problems that actually make people worry about personal identity in real life, which are mostly problems that arise over internal conflicts and clashes of loyalty to different groups around us. We will come back to these conflicts presently.

The difficulty of talking sense about detachable souls afflicts real, professional science-fiction writers too, for their art is deeply committed to dualism. They often produce transmigrational stories in which characters in a wide range of situations keep jumping into other people's bodies, or having their own bodies taken over by an alien consciousness. It even happens in *Startrek*, which shows how natural the thought still is today. But, in order to be convincing, the authors have to fill in a rich imaginative background that links this situation to normal life. And these stories are still strangely limited because they proceed on such an odd assumption.

They treat soul or consciousness as an alien package radically separate from the body. They go on as if one person's inner life could be lifted out at any time and slotted neatly into the outer life of someone else, much as a battery goes into a torch or a new cartridge into a printer. But our inner lives aren't actually standard articles designed to fit just any outer one in this way. The cobbler's mind needs the cobbler's body. It is not likely that two people with different nerves and different sense organs would perceive colour, sound or touch in the same way, let alone have the same feelings about them, or that their memories could be transferred wholesale to a different brain. Trying to exchange their bodies is not really at all like putting a new cartridge in the printer. It is more like trying to fit the inside of one teapot into the outside of another. And this is something that few of us would attempt.

Ships and Pilots;
Batteries and Torches

It is surely very interesting that so many writers of science fiction have signed up for this strange metaphysic. Of course there is nothing odd about their dealing in metaphysics in the first place. Scifi arises out of metaphysical problems quite as often as it does from those in the physical sciences, and good scifi stories can often be metaphysically helpful. But reliance on this particular metaphysic seems to be part of a rather unfortunate recent trend to simplify the relation between our inner and outer lives by talking as if they were indeed completely separate items. This has the unfortunate effect of making it even harder to connect them sensibly — even harder to see ourselves as a whole — than Descartes' separation of mind and body had already made it. Since his time, dualism has persisted and it has grown a good deal cruder.

It is interesting that Descartes himself did not actually show souls as totally disconnected from bodies. Though he ruled that they were substances of different kinds, he placed them both firmly within the wider system of God's providence. He thought God must have good reasons for connecting them, even though those reasons were obscure to us. In fact, Descartes surprises his reader by saying twice explicitly that the soul or self is NOT actually a loose extra added to the body.

He writes:

> *I am not only lodged in my body as a pilot in a vessel....* I am besides so intimately conjoined, and as it were intermixed with it, that my mind and body compose a certain unity. For if this were not the case, I should not feel pain when my body is hurt (Descartes, 1937, p. 135).

Descartes actually knew quite a lot about nerves. He saw that treating the soul as an alien, arbitrary item raised great difficulties about action and perception, so he assumed some underlying connection. And in this he was in tune with Christian thinking, which insisted on the Resurrection of the Body. Bodies were to be restored along with souls at the Resurrection.

But unfortunately Descartes' occasional statements of this link don't stop him arguing all the rest of the time that the separation is absolute. He identifies his self, his 'I' entirely with the soul, the pure spark of consciousness. He speaks of the body as something outside it, something foreign which the soul discovers when it starts to look around it. (The pilot wakes up, so to speak, and finds himself mysteriously locked into his ship.) Descartes rules that, 'the natures of these two substances are to be held, not only as diverse, but even in some measure as contraries' (ibid., p. 76). They have no intelligible relation. Only God's mysterious plan can hold them together.

A soul conceived in this way is, of course, well-fitted to survive on its own after death, which is something that concerned Descartes. It could travel well. *But immortality is not the first thing we need to consider when we form our conception of ourselves.* Before we fit our minds out for the afterlife we need, first and foremost, to have a view of them that makes good sense for the life that we have to live now. By making them so thin and detachable as to be thus independent, Descartes put our inner lives in danger of looking unnecessary.

As the Enlightenment marched on and God gradually faded into the background, the enclosing framework of providence was lost, while the conviction of a gap between soul and body remained and hardened. Increasingly, the advance of physical science made matter seem intelligible on its own. Mind and body did indeed start to look more like ship and pilot. And then, starting from that picture, people began to wonder whether the pilot was actually needed at all. If perception and action were physical processes that could go on without him, had he any function?

These were the thoughts that led the behaviourist psychologists to drop him overboard, leaving a strictly material world of self-directing ships — uninhabited bodies. Descartes' theistic dualism turned into materialistic monism. Subjective experience was dismissed as an ineffective extra, a mere by-product, irrelevant froth on the surface of physical reality. That is why, for a time, people who wanted to seem scientific were not allowed to mention their own or anybody else's inner experience.

But it is very hard to discuss human life intelligibly if you have to ignore most of its more pressing characteristics. Even the most docile of academics don't obey these vetoes for ever. So, as we have seen, eventually some bold people who had noticed that they had inner lives suggested that there was after all this problem of consciousness. (Apparently, it was just one problem...) And now everybody wants to talk about it. But it is notably hard to do so.

One thing that makes the difficulty worse is that scientifically-minded people tend to see this 'problem of consciousness' as a problem of how to insert a single extra term — consciousness — into the existing pattern of the physical sciences and handle it with methods that are already recognised there as scientific. Thus, the famous Tucson conferences on the subject say that their goal is to produce, not an understanding of consciousness but 'a science of consciousness', which it is presumably hoped would be just one more scientific speciality, perhaps something comparable with the sciences of particular kinds of material?

This project is an attempt to revive Descartes' highly abstract soul — his pure spark of consciousness — and to fit it in somehow within the study of the physical world. Since the whole point of separating off this soul-concept in the first place was that it couldn't be handled by the methods used on the physical world, this can't work. Descartes was right about that. What we need now instead is to stand back and consider human beings quite differently — not as loose combinations of two incompatible parts but as whole complex creatures with many aspects that have to be thought about in different ways. Mind and body are more like shape and size than they are like ice and fire, or oil and water. Being conscious is not, as Descartes thought, a queer extra kind of stuff in the world. It is just one of the things that we do.

To grasp this, we need to start by abandoning both the extreme abstractions that have reigned on the two sides of the divide so far.

Inner Lives Are Neither Simple Nor Solitary

At the mental end, we need to get right away from Descartes' idea that the inner life is essentially a simple thing, a unified, unchanging entity, an abstract point of consciousness. He put this point strongly. Unlike body, which is always divisible, mind — he says —

> cannot be conceived except as indivisible. For we are not able to conceive the half of a mind, as we can of any body, however small... (p. 76). When I consider myself as a mind, that

is, when I consider myself only in so far as I am a thinking thing, I can distinguish in myself no parts, but I very clearly discern that *I am somewhat absolutely one and entire.*

... although all the accidents of the mind be changed — although, for example, it think certain things, will others and perceive others — *the mind itself does not vary* with these changes (Descartes, 1937, pp. 76, 139, 77).

This story abstracts entirely from the inner complexity, conflict and change that are primary elements in all subjective experience.

Locke's discussion shows well how misleading this abstraction is. Locke did not dismiss the idea of a separable self or soul, but he asked what it would have to be like if it did exist. He was intrigued by the idea of reincarnation because he had (it seems) a friend who claimed to have been Socrates in a former life. So he asked what we would say if we did come across a case like this where the familiar whole seemed to be divided.

Is the transmogrified prince still the same person? Yes he is, said Locke, provided that he keeps his memories. The word *person* is, he says, essentially 'a forensic term', one centring on responsibility, and we are only responsible for what we can remember doing. With continuity of memory you can still be called 'the same person'. But if you now have a different body, you can't be called 'the same man'.

This suggestion notoriously led to further muddles. But Locke was surely right that any usable idea of a self or person does have to be the idea of something complex, and also of something socially connected with the surrounding world. It must be an entity that incorporates the whole content of a life, the richness of a highly contingent individual experience. Even within the restricted forensic context he at once sees this need for complexity because of its bearing on justice. What (he asks) is to happen if an offender really has no continuity of memory? In that case — he says —

the same man would at different times make different persons; which, we see, is the sense of mankind in the solemnest declaration of their opinion, human laws not punishing the mad man for the sober man's actions, nor the sober man for what the mad man did, thereby making them two persons; which is somewhat explained by our way of speaking in English, when *we say, such an one was 'not himself' or is 'beside him-self'* (Locke, 1924, Bk. 2, Ch. 27, Sec. 15; emphasis mine).

If He Was Not Himself, Then Who Was He?

It seems that after all people are not simple unities, they are highly complex items often riven by inner conflict. Even the law, which usually ignores these complications, cannot always do so and in ordinary life they are matters of the first importance. We often have to consider, not just 'is this man in the dock the same person?' but 'am I myself altogether the same person? Am I (for instance) really committed to my present project?' or again 'which of us within here should take over now?' There are law-courts inside us as well as out in the world. A friend of mine used to say that he unfortunately contained a committee. The trouble was not just that the members didn't always agree but that, when they disagreed, all too often the wrong person got up and spoke all the same.

The truth is that the unity of a human being is not something simple, not something given. It is something terribly difficult, a continuous ongoing project, an aim that has to be continuously struggled for and is never fully attained. Carl Jung called it 'the integration of the personality' (individuation) and thought it was the central business of our lives.

The Importance of Conflict

Plato, who was a very different kind of dualist from Descartes, thought so too and gave conflicts of this sort a central place in his theory. These conflicts take place (he said) within the soul itself and they are a torment to it. The soul is by no means a unity. It is divided into three parts — good desires, bad desires and Reason, which is the unlucky charioteer trying desperately to drive this mixed team of horses.[1] This is, of course, primarily a moral doctrine. But it is also an integral part of Plato's metaphysic and its psychological acuteness has been widely recognised. Its difference from Descartes' scheme shows plainly that *there is not just one way of dividing up a human being.*

There is no single perforated line marking off soul from body, no fixed point at which we should tear if we want to separate

---

[1]   See Plato, *Phaedrus*, sections 246–257.

them. Many ways of thinking about this are possible. None of them is specially 'scientific'. Each is designed to bring out the importance of some particular aspect of our life. Plato's main concern was with emotional conflicts within the self, notably those that surround sex. Descartes, by contrast, was most disturbed about an intellectual conflict, one that arose between two different styles of thinking. It is not surprising that these different biases led them to different views about what a person essentially is. But something that they have in common, and which we may want to question, is that they both wanted to settle the matter by finding one ultimate arbitrator — by crowning one part of the personality as an absolute ruler and calling it Reason.

Just as Hobbes, in trying to end political feuds, put all his trust in a single absolute sovereign, so these moralists, in discussing the feuds within us, want to appoint an inner monarch against whom there is no appeal. They aren't prepared to leave decisions in the hands of a committee. And plenty of people have tried this. But it has never been altogether satisfactory. Today, many of us may think that, although the committee system gives us a great deal of trouble, it is perhaps the least bad alternative that is available to us.

Once we notice this inner complexity we begin to see that it makes the solipsistic isolation of the simple 'thinking thing' impossible too. Inner complexity echoes, and is linked to, a corresponding complexity in the world around us. The divided self is not an independent unit, quarantined from outside interference. Wider patterns outside affect its structure. As Locke saw, a person who has a memory must be an active social being, one capable of being involved in responsibility. Our personal identity is shaped by the surrounding world, depending radically on the attitudes of others.

Thus, when King Lear's daughters begin to treat him disrespectfully, he first says to Goneril, 'Are you our daughter?' and then,

> Doth any here know me? Why, this is not Lear;
> Doth Lear walk thus? Speak thus? Where are his eyes?
> Either his notion weakens, his discernings
> Are lethargied — Ha, waking? — 'tis not so.
> Who is it that can tell me who I am?
> (*King Lear*, Act 1. Scene iv, lines 215 and 223-28)

To which the Fool replies, 'Lear's shadow'.

At this point Lear is speaking somewhat sarcastically. But he soon has to confront these questions literally. The whole point of the play is that his identity has so far centred on being treated as a king. He can't see how to exist without it. And though his case is a specially dramatic one, this point about the crucial importance of social context holds for all of us. The role that we play in the social drama has huge force in shaping who we are. No human being exists in the artificial isolation of the Cartesian pure thinker. When Lear asks who he is, it would not help him to be told that he is a thinking thing.

The Price and the Rewards
of Dualism

Descartes supposed himself to be abstracting from all social influences. He thought he had withdrawn into a realm of pure intellect, designing *a priori* an impartial picture of human knowledge. But the most withdrawn thinkers still take the premises of their reasoning into their sanctuary with them. Descartes was in fact responding to certain quite particular pressures of his own time, trying to resolve the doubts and debates that fuelled the fierce religious wars of his day. He hoped to find a system of thought so universal, so compelling that it could accommodate conflicting theological views and also take in physical science, which might soon begin to rival them.

He devised his dualism as a way of fitting that new science into European culture without harming its Christian background. And, because he wanted above all to unify the system — to avoid doubts and divisions within it — he concentrated intensely on the problem of knowledge. He made the assumption, which has turned out not to be a workable one, that by reasoning we can get absolute, infallible certainty for our beliefs. That is why the soul that he pictured turns out to be essentially an intellect, a reasoning and knowing subject rather than an acting or a feeling one. For him the centre of our beings is the scientist within.

For a time this ingenious division of intellectual life did succeed. It suited Newton well enough. For a great part of the eighteenth century, scientists managed to divide themselves internally to suit the two permitted viewpoints. In their work, they could function as pure thinking beings — that is, essentially as mathematicians. They could view the world around them as an abstract moving pattern, a mass of lifeless, inert particles driven ceaselessly here and there by a few simple natural forces. The rest of the time they could respond to it normally as a familiar rich, complex jumble full of living beings who supplied the meaning for each other's lives. A benign God still regulated the relation between the two spheres.

But as time went on and technology advanced, the more abstract, scientific way of thinking gained strength and pervaded people's lives. Inevitably, conflicts between these two approaches were noticed. As the gap between them widened and became more disturbing, it grew hard to treat them as having equal importance — hard not to ask 'but which of these stories is actually the true one? Which tells us what the world is really like?' People felt that this question had to be answered — that one realm must be accepted as genuine and the other demoted to an illusion. They felt this because it seemed that, if both were equally real, there was no intelligible way of connecting them and reality was irremediably split. Hence McGinn's worry about 'a new kind of reality'. Hence the question that disturbs him and many other people:

> If the brain is spatial, being a hunk of matter in space, and the mind is non-spatial, how on earth can the mind arise from the brain?... This seems like a miracle, a rupture in the natural order (p. 115).

Or, as he puts it after citing a lively scifi illustration, 'The point of this parable is to bring out how surprising it is that the squishy gray matter in our heads — our brain-meat — can be the basis and cause of a rich mental life' (p. 8).

But this is an extraordinary abstraction from reality. Brains do not go about being conscious on their own. Meat is, by definition, dead. Conscious pieces of matter are never just consignments of squishy grey matter, sitting on plates in a lab like porridge. They are living, moving, well-guided bodies of animals, going about their business in a biosphere to which they are naturally adapted. And the question about them is simply whether it makes sense to diagnose consciousness as an integral, necessary, appropriate, organic part of entities in this situation — including ourselves — or whether it is more reasonable to suppose that they might all quite as well actually be unconscious.

What Sort of Explanation?

It is important to notice just what we are trying to do here if we want to 'explain consciousness' in a way that resolves McGinn's metaphysical difficulty. The point is not, of course, just to find some physical condition that is always causally conjoined with it. We want to make that junction intelligible — to show that the one item is in some way suitable to the other.

When one is trying to find the connection between two things in this way — for instance the connection between roots and leaves or between eyes and feet — the best approach is not usually to consider these two on their own in isolation. It is to step back and look at the wider context that encloses them. In the case of consciousness that context is, in the first place, organic life and, in the second, the power of movement.

Any being that lives and moves independently, as animals do, clearly needs to guide its own movements. And the more complex the lives of such beings become, the more subtle and varied must be their power of responding to changes that are going on around them, so that they are able to respond flexibly. That increasing power of responding calls for an ever-increasing power to perceive, think and feel. So it necessarily calls for consciousness, which is not an intrusive supernatural extra but as natural and appropriate a response to the challenges that confront active life as the power of flying or swimming. Plants can get on without such a power but animals could not because they are confronted with problems of choice. We ourselves do a lot of things unconsciously — that is, without attention — but when a difficulty crops up and choice is necessary, we rouse ourselves and become conscious of it.

There is no miracle here. The really startling factor in this scene is something which is usually ignored in these discussions, namely the introduction of life itself. Indeed, one might be tempted to say that consciousness is merely the superlative of life — just one more increase in the astonishing power of spontaneous development and adaptation which distinguishes living things from stones. Once life is present, the move from inactive creatures to highly-organised moving animals is simply one more stage in the long, dazzling creative process which is already a kind of miracle on its own, but one which is not usually treated as a scandalous anomaly.

Discontinuities Within

Can it be true that there is not really an alarming gap here? If so, what is it that has made this particular transition seem so strange?

The sense of strangeness arises, I think, simply from *the shift that we have to make in our own point of view when we consider it.* When we are confronted with a conscious being such as a human, all our social faculties at once leap into action. We are sure that it has an inner life. Questions about its thoughts and feelings at once engage our attention. We bring to bear a whole framework of social concepts, a highly sophisticated apparatus that works on quite different principles from the one we would use if we were thinking about squishy grey matter in the lab.

This shift of methods can raise great difficulties, particularly on the many occasions when we need to use both these ways of thinking together. It makes trouble for instance over mental illness. We find it very hard to harmonise our thoughts about the inner and the outer life of disturbed people — again, perhaps including ourselves. We often run into painful confusions. But *the clash in these cases is not a cosmic clash between different forms of reality. It is not a clash between ontological categories in the world, not a clash between natural and supernatural entities. It is a clash between two distinct mental faculties within ourselves, two distinct ways of thinking, along with the various emotional attitudes that underly them. It constantly raises moral questions about how we should act in the world, questions about what is most important in it.*

This discontinuity does not, I think, actually raise metaphysical questions about what is real. But that is not to say that it is a trivial matter — quite the contrary. The difficulty of bringing together the different parts of our own nature so as to act harmoniously is a crucial one in all areas of our lives. The reason why we are so highly conscious is that we are complex social beings and this means that our choices are never likely to become simple.

Matter Is Not Simple Either

As I suggested earlier, the sense of bizarreness infesting the mind-body conjunction is made worse by the extreme abstraction to which both these terms have been subjected. Here I think the parallel with apartheid is actually quite illuminating. 'Black' and 'white' are extremes of the colour-range. If they are colours at all they are colours that are never actually seen on any human skin. The use of this dramatic contrast to categorise the vast range of people found in South Africa or anywhere else imposes a quite irrelevant, artificial way of thinking, an approach which distorts all perceptions of these populations and makes it impossible to understand their diversity realistically. In the same sort of way, the sharp contrast between extreme conceptions of mind and body has obscured our thinking when we try to meditate on the complexities of our nature.

We have seen how, at the mental end of this mind-body axis, the idea of soul or mind became narrowed to a bare point of consciousness. But at the other end of it too the idea of matter has also been narrowed. Indeed, muddles about matter have probably been even more disastrous than muddles about mind.

Under a blindly reductive approach, the conscious animal that we ought to be asking about is reduced to a brain and even the brain loses its structure, becoming just a standard consignment of chemicals — porridge, squishy grey matter-as-such. It was, however, a central doctrine of seventeenth-century dualism that matter-as-such is inert and can do nothing, all activity being due to spirit. That is surely the conviction that still makes people like McGinn feel that a miracle must be involved if something material takes the enterprising step of becoming conscious.

This thesis of the inertness of matter is not often stated explicitly today, but it is often implied. Peter Atkins expressed it strongly in his book *The Creation* (1987) when he made the startling remark, 'Inanimate things are innately simple. That is one more step along the path to the view that *animate things, being innately inanimate, are innately simple too*' (p. 53).

Animate life, Atkins suggests, is not a serious factor in the world. It is just a misleading surface froth that obscures the grand, ultimate simplicity revealed by physics. Life has no bearing on consciousness, which (he explains) appears in the universe independently of it:

> Consciousness is a property of minute patches in the warm surfaces of mild planets.... Here now (and presumably cosmically elsewhere at other times) the patches are merging through the development of communication into a global film of consciousness which may in due course pervade the galaxy and beyond.... Consciousness is simply

> complexity.... Space itself is self-conscious.... Conscious-
> ness is three-dimensional (pp. 71, 73, 83 & 85).

This is scandalously muddled talk. Consciousness is not a
property of such patches but a property that (as far as we know)
is found nowhere in the universe except in certain rather complex living beings — in fact in animals. And that is the only context in which its presence makes sense.

This kind of attempt to make consciousness respectable as an
isolated phenomenon, without mentioning biological considerations, by inserting it directly into physics and treating it mainly
as a basis for cybernetics, the IT revolution and the colonisation
of space is rather prevalent at present. Similarly David Chalmers
suggested that, in order to avoid reducing mind to body, we
should take 'experience itself as a fundamental feature of the
world, *alongside mass, charge and space-time*' (Chalmers, 1995).
This list shows his conviction that, in order to be fundamental, a
feature must belong to physics. He does not name life as one of
these fundamental features, and he goes on to remark with satisfaction that — if this view is right —

> then in some ways a theory of consciousness will have more
> in common with a theory in physics than a theory in biology.
> Biological theories involve no principles that are fundamen-
> tal in this way, so biological theory has a certain complexity
> and messiness about it, but theories in physics, insofar as
> they deal with fundamental principles, aspire to simplicity
> and elegance.

In talk like this, the desire to keep one's theories clean of messy
complications takes precedence over any wish to get a useful
explanation. Such physics-envy is one more consequence of the
unlucky fact that, in the seventeenth century, modern physics
gained huge status because it was invented before the other sciences. This gave the Newtonian vision of the physical world an
absolute standing as the final representation of reality, which is
why that vision is still the background of much thinking today. It
is surely the source of Atkins' amazing contention that all the
things in the world are innately (whatever that may mean)
simple.

That drastic assumption of simplicity was a central part of the
seventeenth century's determination to get final, authoritative
answers to all its questions. Physicists today have learnt better;

190   *Science, Consciousness & Ultimate Reality*

they do not make this assumption. Like other scientists, they still
look for simplicity, but they know they have no right to expect it.
And they have, of course, altogether abandoned the simplistic
doctrine of inert matter. Solid, billiard-ball-like atoms have van-
ished entirely. As Heisenberg pointed out long ago,

> Since mass and energy are, according to the theory of relativ-
> ity, essentially the same concepts, we may say that all ele-
> mentary particles consist of energy. This could be interpreted
> as defining energy as the primary substance of the world....
> With regard to this question *modern physics takes a definite
> stand against the materialism of Democritus* and for Plato and
> the Pythagoreans. The elementary particles are certainly not
> eternal and indestructible units of matter, they can actually
> be transformed into each other (Heisenberg, 1962, pp. 58–9).

In fact, when physicists abandoned the notion of solid particles the word 'materialism' lost its old meaning. Though this
word is still used as a war-cry it is by no means clear what significance it ought to have today. That change in the ontology of
physics is one scientific reason why it is now clear that the notion
of matter as essentially dead stuff — hopelessly alien to conscious life — is mistaken. But an even more obvious reason is, of
course, the Darwinian view of evolution.

We now know that matter, the physical stuff that originally
formed our planet, did in fact develop into the system of living
things that now inhabit its surface, including us and many other
conscious creatures. So, if we are still using a notion of physical
matter that makes it seem incapable of giving rise to consciousness, that notion has to be mistaken. We have to see that the
potentiality for the full richness of life must have been present
right from the start — from the first outpouring of hydrogen
atoms at the big bang. This was not simple stuff doomed for ever
to unchanging inertness. It was able to combine in myriads of
subtle ways that shaped fully active living things. And if it could
perform that startling feat, why should it be surprising if some of
those living things then went on to the further activity of
becoming conscious?

Disowning the Earth

Many people have pointed out that Descartes' notion of the
secluded soul played a part in the rise of individualism by cut-

ting us off from our fellow-humans. But I think less attention has been paid to the way in which it cuts us off from the living world around us. Descartes viewed all non-human animals, equally with plants, as literally unconscious automata. An animal, he said, does not *act*. It is driven. Human bodies too were automata; their only difference from the rest of the machinery was that they were driven by the alien soul set within them. All organisms, along with the planet they inhabited, were merely arrangements of inert matter. Life belonged only to spirit. And though views about consciousness have softened somewhat since his time, the more general idea that the rest of the biosphere is something foreign and decidedly beneath us has not shifted half as far as it should have done.

This idea still centres on the old notion of physical matter as dead, inert and alien to us, and it is worth while to notice here where this notion came from. Though Descartes used it for his purpose of isolating physics, it is not an objective conception demanded by science. It is part of an ideology that was long encouraged by Christian thinking, an ideology that centred on fear and contempt for the earth, which was seen as simply the opposite of Heaven. Human souls were conceived as having their real home in a remote spiritual paradise. Earth was at best a transit-camp, a place of trial through which they must pass. All sorts of nuances in our language still reflect this drama. Thus, the Oxford dictionary gives as the meaning of *earthy* — 'Heavy, gross, material, dull, unrefined ... characteristic of earthly as opposed to heavenly existence.'

Pre-Copernican cosmology set this heaven literally in the sky, beyond the concentric spheres that bore the sun, moon, stars and planets. The earth was held to be merely the dead point in the middle of the system, the midden to which worthless matter that could not move upwards eventually drained. That central position was *not* seen as a sign of importance, as is often said, but as a mark of worthlessness, of distance from the celestial heights that held everything real value. After all, what lay at the centre of Earth itself was Hell.

Because of this contempt and fear of the earth, when Copernicus displaced our planet from its central position, Christian people did *not* feel the humiliation that is often said to have followed that move. There was of course a sense of confusion and insecurity. But human souls still had their celestial citizenship. Their salvation was still essential cosmic business.

This sense of complacent independence from the earth did not die away, as might have been expected, with the decline of confidence in the Christian vision. Secular Westerners who stopped seeing themselves as Christian souls subject to judgment did stop expecting their previous welcome in the sky. But this did not lead them — as one would think it might have done — to conclude that they might be only rather gifted terrestrial animals. Instead, they still managed to see themselves in the terms that Descartes had suggested as pure intellects — detached observers, set above the rest of the physical world to observe and control it. So, when they stopped venerating God, they began instead to venerate themselves as in some sense the supreme beings in the universe — intellectual marvels whose production must have been the real purpose of evolution. This rather surprising position is expressed fully today in the Strong Anthropic Principle, and to some extent by other manifestoes of what is now called Human Exceptionalism.

Human intellect, in fact, now shone out as supreme in isolation from the whole animal background that might have helped to explain it, and from the rest of the biosphere on which it depended. 'The mind' did indeed begin to look like a miracle, a self-supporting phenomenon without a context. As Roy Porter (2003) says, 'In a single intrepid stroke, Descartes had disinherited almost the whole of Creation — all, that is, except the human mind — of the attributes of life, soul and purpose which had infused it since the speculation of Pythagoras and Plato, Aristotle and Galen' (pp. 65–66). The physical universe no longer seemed to be what Plato had called it, a mighty living creature. It was simply a more or less infinite pile of raw material provided for humans to exploit.

That exploitation accordingly went on without much check throughout the Industrial Revolution. The pile of resources did indeed seem infinite. Doubts about this are, of course, beginning to be felt now. But the sense of humans as essentially independent, powerful super-terrestrial beings is still extraordinarily strong.

Some people — apparently quite a lot in the United States — still ground this confidence in the Christian heaven, expecting to be carried off there in chariots when disaster strikes. Others use the sky differently, advertising future desirable residences in outer space rather than in the traditional heaven. And even among people who don't go for either of these scenarios, many are still confident that scientific ingenuity will always resolve our difficulties somehow. The vision of ourselves as essentially invulnerable minds independent of earthly support, colonists whose intellects will get them out of trouble whatever may go wrong, is still amazingly strong

Life and Its Effects

This flattering illusion of human separateness and self-sufficiency is surely the really disastrous legacy still left over from Cartesian dualism. It is closely linked to the idea that physical matter is inert. That idea makes our planet appear as a mere jumble of blindly interacting particles senselessly forming themselves into handy products for us to consume. If we want to move to a more realistic notion of ourselves, we need to have a more realistic conception of what the earth itself actually is — namely a living, working system.

Thirty years back, when James Lovelock first suggested that way of looking at the matter and gave it the name Gaia, orthodox scientists dismissed it, just as, half a century earlier, their predecessors had dismissed the idea of continental drift. In fact these two ideas have quite a lot in common; both of them are too large and too surprising to be easily digested. But since that time the details of Lovelock's story have been filled in and its main outlines are now scientifically accepted.

Earth scientists now agree that the presence of living things on the earth has indeed had a huge influence on its fate, an influence that until then had scarcely been recognised at all. Organisms do not only adapt themselves to the conditions around them. They also change those conditions, and often do so in a way that makes them more favourable to their own continuance. On the earth's surface they seem to have done this on a scale that has made all the difference to their fate. They have worked to regulate the soil, the temperature and the mixture of gases in the atmosphere in a way that has kept conditions tolerable for life throughout the three and a half billion years that it has been present on the planet.

We can see the difference this has made by looking at our two nearest neighbours in the solar system. All three planets were originally very similar. What makes the earth so different from Mars and Venus is that some factor present here has prevented the runaway greenhouse effect that led to extreme temperatures and atmospheric conditions on those other planets, making them for ever uninhabitable. Some factor has stabilised conditions here at a level suitable for life. And it has done this in the face of some quite alarming challenges.

For instance, the earth's ambient temperature has managed to remain within the quite narrow range needed for life ever since life first appeared, although the sun is now 30% hotter than it was at that time. The odds against this happening merely by chance are incalculably huge. Similarly, the many gases of the atmosphere have kept their rich and complex mixture at a level fit to sustain life by a constant series of exchanges among the various organisms present, instead of streaming off into space and leaving a static residue — mainly of carbon dioxide — as happened on Mars and Venus.

When Lovelock thought about this strange contrast, it occurred to him that the mysterious factor responsible had to be the presence of living things themselves. This planet, unlike its neighbours, had produced organisms that were able to act on their environment in a way that made their own continuance possible. Some of the mechanisms by which this stabilising influence works have now been identified, though of course their working still need much more investigation. This contrast between the planets was what Lovelock had in mind when he said that the earth is in some sense a living planet while Mars and Venus are dead ones.

There is nothing superstitious about this proposal. Lovelock remarks:

> When I talk of a living planet, I am not thinking in an animistic way of a planet with sentience.... I think of anything the earth may do, such as regulating the climate, as automatic, not through an act of will, and all of it within the strict bounds of science (Lovelock, 1991, p. 131).

But, as he points out, the fact that the idea is validated by science does not stop it having implications for our wider thinking:

> For me, Gaia is a religious as well as a scientific concept, and
> in both spheres it is manageable.... God and Gaia, theology
> and science, *even physics and biology are not separate but a single
> way of thought* (Lovelock, 1988, pp. 206, 212).

This deliberately intriguing remark raises more issues than we
can deal with now. But the point Lovelock makes about the gap
between physics and biology is sharply relevant to our present
topic. Not only has there been a gap between these sciences,
there is a much wider gap in our general thinking here.

The Mystery Is Within

After the enquiry that we have been making, two questions may
well occur to us. One is, 'Why has the unworkable mind-body
dualism that we have been examining lasted so long?' The other
is, 'Why did scientists studying the earth not notice earlier that
organisms might have causally affected the planet, as well as
vice versa? Why did they take it for granted that life was merely
an inconsequential by-product of inorganic phenomena? Why,
in fact, did biologists and geologists not talk to each other on
these matters until the last few decades, when, to their own surprise, they have suddenly brought themselves together in
departments of Earth Science?'

The answers to these two questions are related. The delay on
both points springs from the difficulty that we have in bringing
together two very different ways of thinking — two sides of our
personality — two aspects of ourselves. When we are dealing
with living creatures we think socially. When we deal with lifeless matter we use a quite different approach. The recent history
of our culture has made us peculiarly determined to abstract
from everything social, making a strenuous effort to avoid per-
sonifying the inorganic world. These two approaches call out
different faculties within us and the relations between these faculties are not at all simple. Indeed they are highly mysterious.

McGinn is quite right to say that there are real mysteries in the
world, matters that we are not at all well equipped to understand. Foremost among these mysteries are those that concern
the inner structure of our own minds. We are not totally helpless
here. We can make some sense of this structure if we attend to it
carefully. But if, instead of attending to it, we simply project its

conflicts onto the outer world and attack them there by meta-physical conjuring, we shall get nowhere.

References

Atkins, Peter (1987), *The Creation*. Oxford & San Francisco: W.H. Freeman.

Chalmers, D.J. (1995), 'Facing up to the problem of consciousness', *Journal of Consciousness Studies*, **2** (3), pp. 200–219.

Descartes, René (1937), *Meditations On The First Philosophy*, tr.John Veitch. London: Everyman's Library, Dent & Dutton.

Heisenberg, Werner (1962), *Physics and Philosophy*. New York: Harper & Row.

Locke, John (1924), *Essay Concerning Human Understanding*. Oxford: Clarendon Press.

Lovelock, James (1988), *The Ages of Gaia*. Oxford: Oxford University Press.

Lovelock, James (1991), *Gaia: The Practical Science of Planetary Medicine* London: Gaia Books Ltd.

McGinn, Colin (1999), *The Mysterious Flame: Conscious Minds in a Material World*. New York: Basic Books.

Porter, Roy (2003), *Flesh in the Age of Reason*. London: Penguin.

Alan Torrance

Theism, Naturalism &
Cognitive Science
Can the Academy Make Sense of Itself?[1]

Introduction

It is commonplace to draw attention to the challenges which the naturalistic assumptions characteristic of much contemporary science pose to Christian theism. What is less commonly noted is the extent to which naturalism threatens the fundamental suppositions definitive of the contemporary academy and, furthermore, its incompatibility with the affiliations of large sections of the humanities. At the heart of the quandaries that emerge lies the question of how we correlate certain presuppositions in the natural sciences with wider presuppositions as to what it is to be human. The very existence of the academy assumes that we are essentially responsible beings and that we possess free-thinking intellects which can be concerned with truth.

Consideration of developments in the neurosciences may serve to locate precisely where the focus of these challenges is to be found. It may also serve to clarify the confusions inherent in certain attempts to discount theistic approaches within the academy.

It is no exaggeration to suggest that there is a crisis at the heart of the modern university. Contemporary academia seems to be

---

[1] This paper replicates and integrates material contained in three other papers being edited for publication in three other contexts, notably, two volumes being currently edited by Professor Malcolm Jeeves, to whom I am immensely indebted for guidance and encouragement in this field.

defined, indeed, driven by two mutually incompatible kinds of approach. On the one hand, we have 'naturalism' which is dominant in the natural sciences. And on the other we have a very different philosophy undergirding and sustaining large areas of the humanities, namely, what Alvin Plantinga refers to as 'Enlightenment humanism' or 'creative anti-realism'. We could also call it 'social constructionism'. This family of philosophies is almost as influential in the humanities as naturalism is in the sciences. It is found in history, literary studies, film studies, continental philosophy, gender studies, religious studies and various other departments within the 'social sciences'. Unfortunately, like most dysfunctional affiliations, it is alive and well in many theology departments as well!

Plantinga's Exposé of Naturalism

The person who has explored the tension between these to greatest effect is Alvin Plantinga and it is largely on his writings that I shall draw in framing this discussion. What exactly is meant by 'naturalism'? Roger Trigg (2001) defines it as the view that 'reality is wholly accessible (at least in principle) to the natural sciences. Nothing ... can exist beyond their reach.' (p. 149) For Plantinga (1989), 'Perennial Naturalism', as he calls it, is the view that 'there is no God, and we human beings are insignificant parts of a giant cosmic machine that proceeds in majestic indifference to us, our hopes and aspirations, our needs and desires, our sense of fairness, or fittingness' (pp. 9–10) — the blind watchmaker. As he points out, there is nothing new about this approach. It is as old as Lucretius but has become explicit in much modern philosophy (Dewey, Quine, Davidson, *et al.*) and has become the 'given' assumption of research in the biological sciences, anthropology, socio-biology and the cognitive sciences. Richard Dawkins and Daniel Dennett are two of its better-known advocates. Its effect, Plantinga argues, is to collapse the richness and diversity of reality to one level of interpretation — thereby anaesthetising the academy against the notion of any objective sphere of value[2] or of rights or of human dignity — and,

---

[2]   Plantinga (2000) discusses Herbert Simon's (1990) analysis of altruism. One of the puzzles which Simon (a leading academic cognitive scientist, expert in cybernetics and winner of a Nobel Prize in Economic Science) sought to address is why certain people do not behave in the ways in

ultimately, indeed, against virtues such as truthfulness or academic integrity (cf. Plantinga, 1989).

More relevant to our purposes here, however, is the fact that it raises highly problematic questions vis-à-vis the place of truth within a university — not least, the very status of scientific truth-claims. There are at least two reasons for this. First, truth, to the extent that it involves what one might term accountability to Reality, only has a place within the kind of universe which naturalists envisage or assume *to the extent that* it has an evolutionary function — that is, to the extent that it is part of the species' striving for survival. Naturalism cannot acknowledge any place for truth or, indeed, truthfulness *for its own sake*. To assume that someone was interested or concerned for truth *per se* would be to make reference to a system of values beyond utility or fitness for which 'naturalism', conceived as a world-view, allows no place. Second, naturalism struggles to account for the reliability of our cognitive faculties when they seek to make claims at the metaphysical or philosophical level — and thus for the claims it makes itself. Evolutionary naturalism recognises one all-controlling, governing principle, namely, that the evolution of all our faculties and abilities is driven by adaptation for the sake of success or fitness. This implies that our intellectual capacities will only be likely to develop in such a way that they 'aim at truth' in so far as and to the extent that that 'aiming at truth' is integral to the dissemination of one's genes. This, and its consequence, is tidily articulated by Patricia Churchland, who argues that the most important thing we need to recognise to make sense of the human brain is the fact that it has evolved. She comments,

> Boiled down to essentials, a nervous system enables the organism to succeed in the four F's: feeding, fleeing, fighting,

---

which evolutionary theory would dictate, namely, the effective spreading of one's genes. How, for example, do we explain the Mother Theresas of this world? His answer is framed in terms of two principles, namely (a) docility — some people are docile and tend to do what they are encouraged to do (that is, they do what their peers tell them to do without adequately questioning it!); and (b) limited rationality, namely, as Plantinga explains 'stupidity'! (p. 214, fn 21) In terms of his naturalistic account, therefore, morally virtuous or self-denying people like Mother Theresa are an unhappy quirk of fate. Happily, their insufficiently evolved and evolving genes will be condemned by the evolutionary process and the docile gullibility and stupidity constitutive of their altruism purified from the gene pool!

and reproducing. The principle chore of nervous systems is to get the body parts where they should be in order that the organism may survive.... Improvements in sensorimotor control confer an evolutionary advantage: a fancier style of representing is advantageous *so long as it is geared to the organism's way of life and enhances the organism's chances of survival* [Churchland's emphasis].

She then adds, significantly, 'Truth, whatever that is, definitely takes the hindmost.' (Churchland, 1987, p. 548.)

The problem is illustrated when naturalists like Dawkins, for example, make truth-claims of a metaphysical kind. Despite his disdain for theology, Dawkins has no qualms about speculating about what God would or would not do if he existed and even fewer qualms about making absolute metaphysical claims regarding the existence (i.e. non-existence) of God. (See Poole, 1994, 1995; Dawkins, 1995.) This raises the question as to whether these claims (and the supposition that they are aiming at 'truth') do not require to be justified, in evolutionary terms, by way of their contribution (or the contribution of their grounds) to human fitness. If they can be justified with recourse to the over-arching criterion definitive of human emergence and development, (namely, contribution to human fitness), the further question concerns whether these claims might also be regarded as *true* as opposed to merely beneficial or useful?[3] Could it not be the case that proper evolutionary adaptation would involve a propensity to radical self-deceit in certain fields?

What grounds could Dawkins or Dennett offer for claiming that their own naturalistic claims are not to be conceived as the unwarranted projections of social mechanisms which testify to little more than an *un*trustworthy psychological desire to offer empiricist or reductionist accounts of reality — albeit one that serves the dissemination of one's genes? If too much or, indeed, too little belief in a God (defined in one way or another) impacts on the dissemination of our genes then we are justified in assuming, on such accounts, that the human race will ultimately find itself genetically programmed either to believe in such a God or not to do so. The truth or falsity of the claim must surely be

---

[3]   Cf. Alvin Plantinga's evolutionary argument against naturalism in *Warrant and Proper Function* (1993), chapter 12, esp. pp. 228–237.

regarded as an irrelevance in that it buys into an extrinsic philosophy of truth and morality without any clear fitness-related warrant. Ironically even the business of assessing whether it is an irrelevance or not must *itself* be determined by our programmed orientation to say or discover or perceive whatever is most likely, either by a direct or indirect route, to promote 'fitness'.

In short, evolutionary naturalists must acknowledge the possibility of a naturalistically conceived mechanism which leads them to oppose ethical or religious constraints which are likely to constrain the dissemination of our genes. Just such a mechanism might take the form of a desire to repudiate the existence of a higher authority sustaining personal values and obligations which might promote sexual restraint. If so, Dawkins' theory might well establish that he has indeed evolved to the point where he is genetically programmed to repudiate the existence of God. Those who do not repudiate the existence of God will, of course, be less than fully evolved and their genes will be cast aside — along, of course, with the capacity to engage objectively with the relevant truth questions!

The question which all of this raises is whether naturalism could ever succeed in providing warrant for the existence of the academy or does the existence of the latter not presuppose the kind of moral universe that naturalism precludes? Is there any warrant for the existence of the academy without the concept of truth and truth-telling on the one hand and without recognising the inherent intelligibility of the natural order on the other? It is far from clear that naturalism can provide either.

Social Constructionist Approaches

We must now, however, turn to the other philosophical allegiance shaping the university. If 'naturalism' is widely influential in the sciences (as these include the cognitive sciences) and appears to play down the role of human creativity and free agency, the second way of thinking about the world and our role within it, is non-existent in the natural sciences — but widely influential in the humanities and the social sciences.[4] This is

---

[4]   Cf the spectacular hoax pulled off by Alan Sokal, professor of physics in New York, in *Social Text* (one of America's leading journals of cultural studies), by writing a spoof, social constructionist critique of the scientific method, titled, 'Transgressing the Boundaries: Toward a Transformative

what Plantinga describes as 'Enlightenment humanism' or 'creative anti-realism'. That wedding of Kant and Wittgenstein which has generated 'social constructionist' approaches belongs to this genus of philosophy as well. Stemming from Immanuel Kant, 'creative anti-realism' (father of so-called Post-Modernism) suggests that we human beings are fundamentally responsible for creating the structure and nature of the world — it is *we* who are ultimately the architects of any apparent rationality in the universe (Plantinga, 1989, p. 14). For Immanuel Kant, of course, there remained the '*Ding-an-sich*' — the 'thing-in-itself' — which transcended the business of human construction in accordance with the laws and principles of thought and understanding. Significantly — and, indeed, consistent to the thrust of this approach — this concept was dropped by the Marburg Neo-Kantians (Cohen, Natorp, *et al.*), for whom the 'thing-in-itself' must *itself* be interpreted as a feature of the subjective realm, namely, as the concept of a limit on our construction of reality. Any such notion can and must, they suggested, be given adequate explanation in terms of the laws of thought and reason. What emerges is a kind of radical subjectivism — though not one that suggests that this 'creative activity' is free or arbitrary. The construction of objects (*Objektivierung*) is carried out in accordance with universal laws of thought. So-called Postmodernism pushes further and questions the need to suppose any such universality, thereby generating the anti-realist mood which characterises so much of what goes on in the humanities today.

So what are the ramifications of this position for the concept of truth? Richard Rorty appears to see truth as denoting nothing more than 'what our peers will let us get away with saying' (Rorty, 1979, p. 176). For Don Cupitt, it is 'the state of play'. In short, 'if everyone believes x, x is true'. For social constructionism or creative anti-realism, there is no objective 'fact of the matter', there is simply the plethora of ways in which communities divide up and construct the world by means of socially grounded concepts and projected 'meanings'. The relevant human habits

---

Hermeneutics of Quantum Gravity", *Social Text*, **46/47**, pp. 217–252 (spring/summer 1996). Sokal revealed his parody in 'A Physicist Experiments with Cultural Studies' in *Lingua Franca*, May/June 1996, pp. 62–64. See also his explanations of why he wrote the parody in *Dissent* **43** (4), pp. 93–99 (Fall 1996) and *Philosophy and Literature* **20** (2), pp. 338–346 (October 1996).

of thought constitute a world which is not merely inseparable from but identical with these constructive processes.

So wherein the crisis? This lies in the fact that the contemporary academy can now be seen to be composed of two massively influential but radically mutually incompatible ideologies or intellectual commitments. The suppositions of so much that takes place in the humanities are in irreducible and irresolvable conflict with all that is going on in the natural sciences. Both domains are sustained within the modern 'Universitas'! It is far from clear that either of these is able to offer an internally consistent account of the nature of mind or the function of truth within the academic enterprise. It is demonstrably the case that neither is able to provide an account that is consistent with the other! Ironically, however, what would appear to be the most widely shared ideological commitment of both groups is a shared opposition to theistic approaches within the academy.

The irony of this is that theism *can* and *does* offer a cogent and coherent account of these and, consequently, offers integrated justification for the existence of the *universitas*. Theists believe in truth, that there is a 'fact of the matter' which is not contingent upon the subjective state of their minds. They also believe that a fundamental purpose and function of human beings is to seek to access it and, indeed, correspond to it in their lives. Moreover, they have grounds for affirming the intelligibility of the contingent order. They believe in a sphere of value, of which 'truth' is part. They believe that there are forms of activity which are intrinsically wrong, such as lying, the fabrication of academic results whether or not one's peers are in a position to discover their doing this. Moreover, they have grounds for explaining why there is something rather than nothing — an issue which Dennett and Dawkins cannot begin to explain and which surely impugns the latter's claim that Darwin made it possible to be an intellectually fulfilled atheist.

In short, if theism is a little out of fashion in sections of the contemporary university, theists operate with an epistemic base which has, comparatively, a spectacular degree of explanatory power as well as internal coherence! In sum, it is ironic that, despite the fact that naturalists and non-realists all too often argue that theists are obliged to justify their existence in the modern academy, they alone can offer an integrative account which makes sense of the nature, purpose and function of the academy and its fundamental convictions and commitments. In short, they affirm the fundamental tenets without which commitment to the academic enterprise cannot be sustained: the intelligibility of the universe, the heuristic capacity of the mind to access this, the dignity of humanity (and the values of the academy associated with this), the freedom of the intellect and the propriety of pursuing truth for its own sake!

The Cognitive Sciences and the Need to Span the Chasm?

One of the apparent weaknesses of theism in the past has been its tendency to adopt a kind of dualism which fails, among other things, to take account of the profound explanatory success of so much of what goes in the hard sciences — as these include the neurosciences. The question I should now like to consider concerns whether it is possible to approach the nature of mind in interaction with the neurosciences in such a way as to open ways of thinking about human beings which obviate the reductionistic tendencies of naturalism and allow us to affirm the fundamental ideals presupposed by the humanities. Naturalism has tended to favour materialist or physicalist accounts of the person. The appeal of such approaches has been bolstered by the profound problems associated with dualist accounts in the Cartesian tradition so widely appealed to by theists.

The danger with naturalistic accounts is that they risk reducing the whole sphere of human thought and consciousness (and thus the totality of our reasoning and, indeed, thinking about truth, beauty and morality) to a series of epiphenomena explicable exclusively in terms of neuro-mechanical processes. What this entails is that, whereas we assume our thinking about such matters has a degree of autonomy possessing the ability, among other things, to follow conceptual paths, all these mental events require to be conceived in electrochemical terms as a complex causal series of synaptic firings and the like. Given that these events are accounted for in radically physical and, apparently, impersonal terms, it is difficult to see how the path of conceptual reasoning can be identified with a series of physical causal interactions between electrochemical brain-states.

Why Physicalist Accounts of the Mind Are Inevitably Reductive: Jaegwon Kim's Repudiation of 'Non-Reductive Physicalism'

Given the problems associated with dualism on the one hand and its apparent failure to take sufficient account of the physicality of our cerebral operations, a number of thinkers in recent years (including theistic philosophers) have advocated a position referred to as 'non-reductive physicalism'. The hope is that this might preserve human subjectivity while denying a reductionist agenda. This, it might be suggested, is necessary to make sense of the freedom and those other facets of human subjectivity which are assumed by the kind of 'detached objectivity' and truth- seeking presupposed by the academic enterprise.

'Non-reductive physicalism' sets out to hold two theses simultaneously. On the one hand, it holds that thoughts, intentions, fears, etc., cannot be reduced to physical properties — thinking about whether Arsenal will win the league is not reducible to a physical property. This 'neural firing has the following physical properties … including that of "being a thought about whether Arsenal will win the league"'. On the other hand, however, it holds that all causality is nonetheless physical — takes place between purely physical entities.[5] The appeal of such a view is that it appears to combine the strengths of both positions. This position was vigorously attacked by Jaegwon Kim who argues, in his book *Supervenience and the Mind* (1993), that one can adopt either a physicalist approach or a non-reductivist one but not both (pp. 351–352). This he explains with reference to the following simple diagram:

M causes M*

P causes P*

The focus of his argument is the suggestion that one mental event (M) causes another mental event (M*). The first mental event might be a thought of the following kind: 'Because I went to Dublin this weekend, my sons missed skiing in Glenshee.' The second mental event (M*) is a thought caused directly by the former, for example, 'I shall take the kids skiing in Glenshee next

[5] This section draws extensively on Teed Rockwell's particularly, perceptive analysis and summary of Kim's position in his article, 'Non-reductive Physicalism' in the online *Dictionary of the Philosophy of Mind*, ed. Chris Eliasmith (no date).

weekend.' The second thought resulted from the first. The first thought brought about, that is, it *caused*, the second.

If you are a non-reductive physicalist then you must regard the mental event or 'thought' M as 'physically realised' in the form of a physical brain state 'P'. Physicalists must then interpret 'P' as causing brain state 'P*', that is, the physical realisation of the mental event 'M*'. What Kim shows conclusively, it seems, is that 'M causes M*' becomes superfluous on this account — it contributes nothing whatsoever to the account! Why? Because P can cause P* all by itself, making M entirely redundant. No help from M is required. Furthermore, it is not possible. M cannot cause M* without P causing P*. Put simply, on *all* physicalist accounts, physical causality is all there is, and 'mental descriptions' as Rockwell puts it, 'are somewhere between being shallow and being outright falsehoods.'

This places the so-called non-reductive physicalist on the horns of a dilemma. To suggest that one thought causes or actively leads to another is non-reductive but clearly not physicalist. To suggest that all causality is physical is certainly physicalist but it is also inherently reductive. There just isn't such a position as 'non-reductive physicalism'. You can't have your cake and eat it!

Problems with 'Piggy Back' Approaches to Connections between Mental Events

Kim's argument against non-reductive physicalism is difficult to refute. It also serves to direct us to where the really substantive issues lie. These, as we shall see, concern how we understand the nature of physical or causal explanations *per se*.

It is when we consider this that we begin to find problems with Kim's alternative physicalist position. Kim takes the view that all higher level causation '*supervenes*' upon more basic physical explanations.[6] That is to say, all higher level causation (for example, the kinds of causation we associate with mental events) rides 'piggy-back', to use Loewer's phrase, on more basic causal interactions (Loewer, 1998, p. 310). When we think that one complex entity is having a direct causal impact on another, this is only

---

[6] Explaining the meaning of supervenience, Kim (1998) writes, 'mind-body supervenience is the thesis that any two things, or events, that are exactly alike in all physical respects cannot different in mental respects' (p. 148).

apparent. This 'apparent' causality disguises the source of the real causal interaction. This belongs exclusively to the simplest, the most basic physical constituents of these entities. But Kim's advocacy here of 'supervenient causation' presents problems. As Rockwell enquires, how does he not end up denying causality to anything other than the most elementary physical components? Kim, indeed, seems to suggest precisely this when he writes, 'all causal relations involving observable phenomena — all causal relations from daily experience — are cases of epiphenomenal causation.' (Kim, 1998, p. 96.)

The implication of Kim's position, therefore, is that no emergent causation of any kind can exist at all. No emergent structures, processes or patterns can have any direct causal effect on anything whatsoever since they are merely 'epiphenomena'. Whereas complex individuals, be they human agents, animals, flies or billiard balls may appear to have causal efficacy, in actual fact they have none. No complex individual can have any kind of capacity for 'downward causation'. All causality is completely and entirely explicable in terms of causal relations between the most basic, sub-atomic components.[7]

In short, physicalism inevitably ends up reducing all causality to the nexus of causal relationships between absolutely basal or microscopic entities — whatever these might be. All causal processes which are assumed to take place at a higher level are merely 'apparent'. When Joanna Bloggs records her dissent over the appointment of Freddie Smith to the chair of metaphysics because his work assumes that counter-factuals of freedom have truth-values, the only *real* causal events taking place involving all the characters, their thought processes and interactions are those basic causal connections between all the individual sub-atomic particles (that is, quarks ... or whatever quarks are

---

[7] Rockwell points out: 'If we are to be consistent in our denial of emergent processes, we must claim that strictly speaking the rock thrown at the chair did not cause the chair to fall over. Rather the relationship between the thrown rock and the chair is an epiphenomenon that supervenes on the genuine causal processes of subatomic particles, in essentially the same way that mental states supervene upon physical states. For if we granted the existence of emergent macroscopic causal properties within physics, there would be no reason to deny their existence in the mental realm.' (Rockwell, no date. As this article is located in an electronic dictionary, page references do not apply.)

composed of)[8] which constitute Joanna Bloggs together with her deductive thought processes, beliefs, arguments and intentional actions and the nexus of sub-atomic particles which constitute those of Freddie Smith. In fact, we are really talking about myriad interactions between sub-atomic particles *which actually recognise no boundaries whatsoever* between the particles constitutive of Bloggs, her colleagues, Smith and their shared physical environment. Causality on this account does not recognise the boundaries of individual entities and, consequently, doesn't recognise individuals *of any kind* be they conscious or non-conscious.

In sum, the thought of Joanna Bloggs regarding the impropriety of believing that there can be true or false claims regarding counter-factuals of freedom can, in and of itself, have no causal impact whatsoever on whether she records her dissent or not! All such mental events are epiphenomena which require to be interpreted exclusively in terms of causal laws applying to the relevant subatomic particles constitutive of the apparently (but *only* apparently) diverse and complex entities involved. Clearly, such an account raises a whole host of questions. For example, how can this kind of account make any sense of how the mind can think through complex theoretical arguments where each step along the way leads conceptually or deductively to the next? How can such conceptual links be accounted for adequately, when explained exclusively with recourse to causal relationships between basic physical particles? More importantly, is it really *scientific* to demand this? If so, what conceivable groundsmight there be for such a requirement? And could those grounds make any sense of themselves?

A Third Alternative
to Physicalism and Cartesianism?

In 1999 the influential philosopher of science, Nancy Cartwright, wrote a book on the philosophy of science, *The Dappled World*, which promises some significant ways forward. The argument of the book suggests an interesting alternative to dualism and physicalism, which is not discussed by Kim, namely, 'pluralism'.

---

[8]   Rockwell likens Kim's requirement that causal entities require us to penetrate beneath all macroscopic 'epiphenomenal' patterns to basal causal entities as being like peeling away endless layers of onion skin.

This is the view that patterns emerge in physical processes which have genuine causal powers. This indicates that there is indeed an 'emergent causality' which derives from physical processes. This in turn suggests a highly complex world — what she describes as a 'dappled world', namely, a variegated universe characterised by a 'patchwork of laws' (Cartwright, 1999, p. 25). Repudiating the 'fundamentalist' assumption that 'all facts must belong to one grand scheme' (p. 25) she argues that the world simply cannot be explained in terms of the operations of one single kind of causal law — what she refers to as 'nomological monism' — even when this is disguised by an account of 'supervenience' (pp. 32–33). Rather, it requires to take account of diverse patternings and interactive processes. The ramifications of her account are summarised and developed by Rockwell (no date) as follows:

> There could be a variety of macroscopic patterns having an impact on such a world, some of which would be able to control the particles they were made of, rather than exclusively the other way around. In such a pluralistic universe, there would be no principled reason for denying the possibility of mental causality. Mental processes could be one kind of emergent phenomenon, but not the only one. One could flippantly say that when one asks a pluralist 'are you a dualist' the correct answer is 'yes, at the very least'. Such a view would save mental causation from having to rely on finding something ontologically unique about the mental, and from being tarred with the brush of Cartesian dualism. In a post-Darwinian world, any attempt to grant special abilities to consciousness (especially to human consciousness) is bound to look like special pleading motivated by wishful thinking. If we can get the same result by seeing our mental processes as one of many different kinds of emergent properties, then mental properties would be a much more plausible result of evolutionary processes.

The fundamental question around which this whole debate revolves, therefore, can now be seen to concern *where precisely causal properties are located*. Do they only belong to physical objects, as most philosophical naturalists hold? If so, are they only to be found at the level of microscopic physical entities? Or are they to be found at the level of complex entities, that is, of higher level patternings? If this is the case, might they not also be found at the level of 'minds' or, indeed, 'persons'? As Rockwell points out, 'Kim has shown us that if we are willing to say that causal properties emerge anywhere between quarks and minds, we have no reason to deny causal powers to minds.' Whereas Kim's world-view suggests they are only to be found at the level of quarks, Cartwright articulates a world populated with an irreducible plurality of diverse and complex entities found on a whole variety of different and diverse levels and which possess causal powers. For Cartwright, these are emergent though precisely how they might emerge to have causal power requires to be clarified. From within a theistic epistemic base, they may well be regarded as 'given' just as the fact that anything exists at all rather than nothing is 'given'. That is, a theistic epistemic perspective may be less likely simply to regard this emergence as incidental or coincidental. What is important to note is that to make sense of the academic enterprise, one is required, as Cartwright sees with such clarity, to repudiate fideistic commitments to nomological monism and to acknowledge the complex diversity of the universe and the plurality of the laws that constitute it. This is the *sine qua non* of appreciating the ability of the human intellect to penetrate the intelligibility of the contingent order — which must remain the essential supposition of the natural sciences. This not only removes some of the cruder reasons for dismissing theistic accounts, it also opens the door for recognising features of the contingent order of which neither naturalism nor anti-realism can begin to make sense.

The concern of the first part of this paper was to try and expose some fundamental problems posed for academia by that form of fideism known as 'naturalism'. We suggested that the contemporary university is not committed to such a position — and that its main opponent is not theism! Rather, academia is rent between naturalistic fideisms on the one hand and creative anti-realist fideisms on the other — both of which are irreducibly incompatible with each other.

The further suggestion was then made that both these positions have to overcome very substantial problems if they are to make sense of the academic enterprise and its pursuit of truth — neither of which constitutes a problem for theism. It was also noted that theism has far greater explanatory power than its alternatives. The fact that naturalism is perceived to have any explanatory power at all with respect to the provision of ultimate explanations relies on its being considered to be the only game in town — a position its advocates have sought to secure by way of their dismissive rhetoric (appealing to a kind of quasi-intellectual political correctness) vis-à-vis theism, a tactic which is now all too common in the contemporary university!

It was then pointed out that materialist or physicalist forms of naturalism struggle to make sense of those very cognitive processes presupposed by all academic research – including, ironically, the argumentative processes which lead to their own conclusions. It is incumbent on them to show, therefore, that they are not hoist on their own petard.

Finally, I suggested that the best way to make sense of cognitive science — and, indeed, its attempts to make sense of itself — is to repudiate naturalism's close associate nomological monism and to recognise a complex world which demands to be understood and interpreted in non-reductive ways. Whereas this is not in itself an argument for the truth of theism, it counts as a powerful argument against the kinds of affiliations which lead 'naturalists' and other scientistic fundamentalists to repudiate theism.

Anthropology from a Theistic Epistemic Basis

Although space only allows us to touch briefly on the matter, the need for an alternative anthropological approach is further highlighted by the influential writings of the philosopher, Galen Strawson, who introduces the issue of the self and causality at another level. For Strawson, nothing can be a cause of itself. Every brain state therefore requires to be understood in the light of a previous brain state. To the extent that a brain state is 'brought about', the morally relevant cause of that is to be understood as the intentionality identified as the brain state at T^{-1} and the cause of that intentionality, namely, the previous brain state at T^{-2}... and so on. Consequently, he argues that it makes no logical sense to suggest that 'the buck stops' with the self (cf. Strawson, 2003). His argument, which I have condensed, runs as follows:

1. How one acts when one acts rationally is, necessarily, a function of, or determined by, how one is, mentally speaking.

2. If one is to be truly responsible for how one acts, one must be truly responsible for how one is, mentally speaking.

3. But then one must have chosen to be the way one is, mentally speaking.

4. But one cannot really be said to choose, in a conscious, reasoned fashion, to be the way one is, mentally speaking, in any respect at all, unless one already exists, mentally speaking, already equipped with some principles of choice in the light of which one chooses how to be.

5. But then one would have to be responsible for these principles of choice.

6. But for this to be so one must have chosen them, in a reasoned conscious fashion.

7. But for this to be so one must already have had some principles of choice.

8. True self-determination is logically impossible.
   (Strawson, 1986, pp. 28–29.)

This kind of argument, to the extent that it operates with a strong sense of the free self as *causa sui*, is problematic for those who would defend freedom, not least, a robust notion of responsible academic or intellectual freedom. When such an account is translated into an interpretation of the self as simply a series of physical brain states and where each brain state is understood as the effect of a previous brain state plus external causes, it is difficult to make any sense whatsoever of free, rational and responsible action. The buck never stops. Every brain state simply becomes one brain state in a closed causal series of physical events.

A naturalistic approach to the neurosciences cannot but compound the challenge issued by Strawson. The fundamental question again, concerns how we understand causality with respect to human agency. If Strawson's account is reworked in the light of a nomologically monist account of the causal relations between successive brain states, (conceived electrochemically in terms of synaptic firings and the like) then the challenge he issues to common sense accounts of the intellectual freedom intrinsic to the academic enterprise becomes all but insuperable.

If physicalism is necessarily reductive, as Kim has shown, and as the implications of applying Galen Strawson's arguments make even clearer, and if substance dualism appears to take insufficiently seriously the physicality of what makes us human, Cartwright's 'pluralism' (which recognises a complex plurality of emergent systems with causal powers) may suggest a way of taking neuropsychology seriously while simultaneously taking account of the fact that thoughts and intentions and abstract speculations do indeed have causal power vis-à-vis physical events — not least the neural firings which take place in our brains. This does not, in and of itself, overcome Galen Strawson's arguments, but would appear to open the door to a non-linear account of the causality inherent in personal agency such that one may be able to argue for the self as *causa sui* in a non-linear, and perhaps weaker, sense!

What cannot be denied, however, is that a great deal more requires to be clarified if we are to offer a cogent account of the diverse forms of causal interaction which are deemed to take place between complex entities. If we are ever to offer a grand unified theory of the self, not only do we need to eliminate the profound problems associated with how to understand causality — its scope and *modus operandi* - but we also need to take seriously subjectivity or what we might term 'I-ness' and the absolute non-parallelism which exists between being an 'object' and being an 'I'. In his classic, *Concept of Mind*, Gilbert Ryle pointed out that there is a systematic elusiveness relating to the notion 'I'. That is, in every attempt to grasp the notion 'I', one's I-ness inevitably eludes one's grasping for the simple reason that the quarry is and will always remain the hunter. No matter how hard one tries one can only ever arrive at yesterday's fair. (Ryle, 1973, pp. 186–189.)

In a very different context Jürgen Moltmann offers a related observation about the human self,

> As he tries to get behind things in order to understand them and to make use of them, he finally wants to get behind himself too, in order to understand himself. But because it is himself behind whom he wants to get, he keeps on slipping out of his own grasp, and becomes more of a puzzle to himself.... The more possible answers he has, the more he feels he is in a hall of a thousand mirrors and masks, the more unintelligible he is to himself. (Moltmann, 1974, p. 2.)

The inherent elusiveness not only of human subjectivity, but of *qualia*, (or consciousness), freedom and agency presents one series of problems. The immense philosophical problems involved in offering a satisfactory account of causality and physical laws reflected in the thought of philosophers from David Hume to contemporaries such as Bas Van Fraassen (1980) and Nancy Cartwright (1983) pose another. To cite C.A. Hooker's conclusions in a recent article on Natural Laws, 'There is no uncontroversial theory of laws: all face difficulties. As fast as we scientifically unravel nature's mysteries, so fast does the nature of that understanding become mysterious.' (Hooker, 1998, p. 474.) If this is the case with regard to natural laws in general, the situation is that much more complicated if we are to seek to make sense of the functioning of physical laws vis-à-vis human subjectivity and agency.

In a concluding Appendix to his *Treatise on Human Nature* published in 1738, David Hume wrote poignantly about the problems that consciousness and personal identity posed for his project: 'all my hopes vanish when I come to explain the principles that unite our successive perceptions in our thought or consciousness. I cannot discover any theory which gives me satisfaction on this head.' (Hume, 1970, pp. 330–331.) Approaching three centuries later and despite all the many developments in cosmology, the neurosciences and philosophy, Jaegwon Kim concludes his *Philosophy of Mind* by acknowledging the dilemma which results from the two 'intractable problems' of consciousness and mental causation. He continues,

> It is not happy to end a book with a dilemma, but we should all take it as a challenge, a challenge to find an account of mentality that respects consciousness as a genuine phenomenon that gives us and other sentient beings a special place in the world and that also makes consciousness a casually efficacious factor in the workings of the natural world. The challenge, then is to find out what kind of beings we are and what our place is in the world of nature. (Kim, 1998, p. 237.)

Christian theists must always remain on guard lest they fall into the trap of postulating a 'god of the gaps', that is, where 'god' is defined in such a way as to plug the gaps in our present understanding of things — and where the 'god' we worship becomes the crude product of our attempts to salve our own ignorance. At the same time, it is entirely appropriate for them to affirm that the complexity of this world is irreducible to anything other than the simplest explanation of all, namely, the God who has created persons for personal communion with himself, with each other and, indeed, with all the other sentient beings which contribute to the richness of the world as we know it. Those who would seek a grand, integrated unified theory of what it is to be a person — and thus find the answer to Kim's concluding question — may have to wait until that time when we come to know ourselves, 'even as we are known'. In the meantime one might add, Christian theism does not begin by seeking to create a space for itself within the academy. Rather it begins by addressing the 'who question' to the One in whom it recognises, as Søren Kierkegaard put it, the eternal has entered time and to that kinship which the eternal has established there. (Kierkegaard, date?, pp. 507–508.)

Christian theism does not, therefore, begin from naturalistic premises and then seek to work up to a fuller account of the person. Rather, it begins from a position that recognises that human beings are complex realities, that they are, moreover, not objects but subjects — 'I's' who are free, responsible, personal agents who not only pray, love, think, reason, guess and hypothesise but have the capacity to penetrate heuristically the intelligible structures of the contingent order. They are beings who grow and develop physically, emotionally and intellectually and where the latter seems to be intrinsically connected to the development of its neurological components. It also recognises that we are beings whose emotions can be chemically induced, whose moral consciences appear to be soluble in alcohol and whose memories and intellectual capacities degenerate with physical deterioration. We are beings who die but, as the Christian faith suggests, stand to be raised from death to life by the One who created the totality of the contingent order from nothing. This requires Christian theism to hold that we have an 'identity' or 'I-ness' which can die and be raised. If this task appears a challenging one, it should be clear that it is no more challenging than the task that the secular academy has to face if it is to make sense of itself from within the diverse, plural and too often internally inconsistent premises which define its present shape and character.

References

Cartwright, Nancy (1983), *How the Laws of Physics Lie*. Oxford: Clarendon Press.

Cartwright, Nancy (1999), *The Dappled World : a Study of the Boundaries of Science*. Cambridge.

Churchland, Patricia (1987), *Journal of Philosophy*, LXXXIV, Oct., p. 548. Cited by Plantinga.

Dawkins, Richard (1995), 'Reply to Michael Poole', *Science & Christian Belie.f*, **7** (1), pp.45-50.

Hooker, C.A. (1998), 'Natural laws', *Routledge Encyclopaedia of Philosophy*, Vol. 5 ,ed. Edward Craig. London: Routledge.

Hume, David (1970), *A Treatise of Human Nature*, ed. D.G.C. Macnabb. Collins, Fontana.

Kierkegaard, Søren (date?) *Concluding Unscientific Postscript*,

Kim, Jaegwon (1993), *Supervenience and the Mind*. Cambridge University Press.

Kim, Jaegwon (1998), *Philosophy of Mind* .

Loewer, Barry (1998), 'Mental causation', *Routledge Encyclopaedia of Philosophy*, Vol 6, 310

Moltmann, Jürgen (1974), *Man: Christian Anthropology in the Conflicts of the Present*, trans. John Sturdy. London: SPCK.

Plantinga, Alvin (1989), *The Twin Pillars of Christian Scholarship*, The Stob Lectures of Calvin College and Seminary, booklet published by Calvin College, MI.

Plantinga, Alvin (1993), *Warrant and Proper Function*.

Plantinga, Alvin (2000), *Warranted Christian Belief*. New York: OUP.

Poole, Michael (1994), 'A critique of aspects of the philosophy and theology of Richard Dawkins', *Science and Christian Belief*, **6** (1), pp. 41–59.

Poole, Michael (1995), 'Response to Dawkins', *Science and Christian Belief* **7** (1), pp.51-58.

Rockwell, Teed (no date), 'Non-reductive Physicalism' in the *Dictionary of the Philosophy of Mind*, ed. C. Eliasmith.
http://www.artsci.wustl.edu/~philos/MindDict/nonreductivephysicalism.html

Rorty, Richard (1979), *Philosophy and the Mirror of Nature*. Princeton.

Ryle, Gilbert (1973), *The Concept of Mind*. Penguin (originally published 1949).

Simon, Herbert (1990), 'A mechanism for social selection and successful altruism', *Science* **250** (December), pp 1665ff.

Strawson, Galen (1986), *Freedom and Belief*. Oxford: Clarendon.

Strawson, Galen (2003), 'The Buck Stops – Where?', *The Believer*, March.

Trigg, Roger (2001), *Philosophy Matters*. Blackwell.

Van Fraassen, Bas (1980), *The Scientific Image*. Oxford: Clarendon.

Denis Alexander

Science, Faith &
Human Values

Attitudes to Science

Contemporary western societies are profoundly ambivalent about science. On the one hand science is invested with exaggerated expectations and inflated hopes. The vision is for a high-tech universe in which we manipulate its powers to serve our own ends. At the other extreme a vigorous anti-science lobby perceives science to be the source of all our current woes. Scientists are viewed as dangerous meddlers, wresting secrets from nature that are best left well alone, playing god as they pry into the sequence of the human genome and uncover the fundamental forces that hold the universe together. As C.P. Snow once said: 'Technology brings you great gifts with one hand, and it stabs you in the back with the other'.[1]

The rapid advances in science which are predicted for the twenty-first century, particularly in the biological sciences, will certainly bring increasing pressure to bear on our notions of human identity and value. Scientific advances are continually throwing up questions which science itself is poorly equipped to address. We will need to draw on all the resources we can lay our hands on if we are to maintain human justice, dignity and worth in the face of scientific disciplines, such as the neurosciences and the new genetics, which increasingly lay bare our own biological

---

[1] Quoted in *New Scientist*, 4 October 2003, p. 3.

constitution. It is for this reason that significant amounts of science funding are now routinely being made available to ethicists, philosophers *and* theologians in order to tackle the pressing moral and ethical questions raised by scientific advances. Without serious public understanding, discussion and debate there is a real danger that science will continue to appear threatening and dehumanising to many people.

In this context it is a matter for regret that the wider implications of science are sometimes portrayed by scientists in hyper-reductionist terms as if the knowledge generated by their own particular discipline was the *only* knowledge of any significance. In the immortal words of Jim Watson, he of double-helix fame: 'There are only atoms. Everything else is merely social work' (quoted in Rose, 1988). Or as Peter Atkins from Oxford has robustly expressed the matter: 'Humanity should accept that science has eliminated the justification for believing in cosmic purpose, and that any survival of purpose is inspired solely by sentiment' (Atkins, 1992). Such comments reflect a view of science that, I would argue, goes well beyond science itself, in which an attempt is being made to use the prestige of science to prop up personal ideologies.

Apart from the intrinsic implausibility that scientific data can be used successfully to scale such metaphysical heights, these hyper-reductionist views give the unfortunate impression to the public at large that scientists are hostile to human values. The caricatures of scientists and of science so loved by the media are reinforced rather than challenged by the creed of the arch-reductionists. Mr Spock, the ice-cold rationalist of the starship 'Enterprise', provides one of the more common media images of the scientist — in this case an emotionless half-human with a utilitarian moral sense who is there to solve complicated problems. A third of all horror movies has a scientist as the villain, not a very comforting thought for those of us within the scientific community. And one really wonders whether such images were challenged or reinforced by the words of Richard Dawkins a few years ago as he assured a lecture-hall full of school-children:

> We are machines built by DNA whose purpose is to make more copies of the same DNA.... That is EXACTLY what we are for. We are machines for propagating DNA, and the propagation of DNA is a self-sustaining process. It is every living object's sole reason for living. (Dawkins, 1991.)

Science and Faith

It is of course impossible to talk about science and human values without also talking about faith. In the rosy optimistic glow that marked the end of the nineteenth century, many thought that as science and education spread, religious belief would decline automatically. Now, more than a century later, we know that this expectation was mistaken. For good or for ill, religious belief continues to exert a dominant influence over the great majority of the world's population, 87% of that population currently considering themselves to be 'part of a religion'. Whilst in some technologically advanced areas of the world, such as Europe, the late twentieth century saw a decline in commitment to institutional religion, in the USA, by any criteria the nation which currently leads the world in science, the reverse happened and religion boomed. All the evidence suggests that humankind, taken as a whole, is incurably religious and the somewhat ethnocentric secularisation models propounded by Western European sociologists back in the 1960s — claiming that modernisation was inevitably linked to secularisation, and that all other countries were destined to pass through the same secularising process as they modernised — has now given way to the realisation that all countries (and often particular communities within each country) have their own unique histories.

In 1968 the American sociologist Peter Berger wrote in the *New York Times*: '[By] the twenty-first century, religious believers are likely to be found only in small sects, huddled together to resist a worldwide secular culture'. But in 1996 we find (the same) Peter Berger writing that: 'The assumption that we live in a secularised world is false. The world today, with some exceptions..., is as furiously religious as it ever was, and in some places more so than ever. This means that a whole body of literature written by historians and social scientists ... loosely labeled as "secularisation theory", was essentially mistaken'.

So if humankind is incurably religious, what kind of resources does religion provide for thinking about human values? It is quite impossible to answer that question without first tackling the prior question as to how faith and science relate. Numerous models have been proposed to describe such relationships and I want to outline just four, by no means an exhaustive list. Once we have provided a whistle-stop tour of these four models, I am then going to re-visit them one by one and ask the question in each case: 'What does this or that model have to tell us about the relationship between science and human values?'

The Conflict Thesis

The *first model* is the 'Conflict Thesis' or 'Conflict Model'.[2] As the name suggests it is the idea that science and religion have historically been locked in mortal combat and that today, still, they provide conflicting explanations for the way that the world is. As the tide of scientific knowledge swept in, so the tide of religious knowledge was forced to retreat, until finally science reigned supreme — or so the triumphalist modernist account would have us believe.

Now there is no doubt that this model remains quite widespread in popular culture — the idea lives on in the media, in TV studios and in the pages of popular science magazines. Producers often still think that conflict makes for better radio or TV so will deliberately look for combatants to fill their studios. The unfortunate effect in the case of science and religion is to nurture a model which has long since passed its sell-by date. At the same time the Model is strengthened by that small sub-set of scientists, some of whom I quoted earlier, who wish to maintain that scientific knowledge is the *only* form of reliable knowledge.

Certain sections of the religious community also help to keep the Conflict Thesis alive. For example, the campaigns by creationists in America to ban the teaching of evolution in schools has, not surprisingly, continued to give the impression to the American public at large that there is some kind of hostility between science and religion. It is difficult to avoid the conclusion that the extreme polarities in this murky debate — with the likes of Dawkins at one pole and the creationists at the other — are actually mirror images of each other, both requiring the continued existence of the opposite extreme to maintain their viability.

From an academic perspective the conflict thesis is virtually dead, certainly for historians of science. As Steven Shapin remarks in his recent book *The Scientific Revolution*:

---

[2]   More on the competing models can be found in Alexander (2001) and Alexander & White (2004).

In the late Victorian period it was common to write about 'the
warfare between science and religion' and to presume that
these two bodies of culture must always have been in conflict. However, it has been a very long time since these attitudes have been held by historians of science. (Shapin, 1996,
p. 195.)

The fact of the matter is that today no contemporary historian
of science takes the conflict thesis seriously as an overarching
model to describe the historical relationship between science
and religion. And the reasons for that are not too difficult to see,
because as a matter of fact the interactions between science and
religion have been extraordinarily long and fruitful, but certainly not describable by any one particular model. There is no
need to rehearse here the hundreds of examples of early scientists who saw their faith as feeding directly into a positive attitude towards the scientific enterprise. Many of them were
founders of the disciplines that we still practice today, such as
the astronomers Kepler and Galileo, the chemist Robert Boyle —
who liked to perform his experiments on a Sunday to underline
the fact that his science was part of his worship — the naturalists
John Ray and Linnaeus, Isaac Newton, Michael Faraday, Clerk
Maxwell, and so the list could go on. There seems little doubt
that belief in God as creator and as the lawgiver who guarantees
the reproducibility of the properties of matter played an important role in the emergence of modern science during the course of
the sixteenth and seventeenth centuries (see Hooykaas, 1972;
Russell, 1985; Lindberg & Numbers, 1986; Brooke, 1991; Harrison, 1998; Brooke & Cantor, 1998).

We do not have time here to delve into the roots of the conflict
thesis, a topic tackled at greater length in my *Rebuilding the
Matrix* (2001). Suffice it to say that in France it was the writers of
the Enlightenment period who painted science with a secularising brush and who utilised the resources of science to attack the
domination of the Church in public life — so leaving in French
society a long-term impression that science was for freedom and
egalité, whereas religion was a reactionary force holding back
the advance of knowledge.

In contrast in Britain science and religion were seen as mutually supportive well into the nineteenth century. This relationship was disrupted not by Darwin's *Origin of Species* but, it has
been suggested, as a side-effect of the professionalisation of

science which gathered pace during the second half of the nineteenth century (Russell, 1989).

The conflict thesis has suffered not only at the hands of the revisionist historians of the latter half of the twentieth century, but also at the hands of sociologists. A reasonable inference from the conflict thesis might be that those with religious faith would be less likely to go into science and/or that scientists are less likely to adopt religious faith. The available evidence does not support such a thesis. Data from the USA published in *Nature* in 1997 suggest that within the scientific community the level of belief in a personal God who answers prayer remained virtually unchanged at about 40% during the course of the twentieth century (Larson & Witham, 1997, 1998; MacKenzie Brown, 2003). When the question is loosened to include broader categories of theistic belief, then the percentages go up — the massive survey carried out by the Carnegie Commission in the USA showed that 55% of those involved in the physical and life sciences described themselves as religious, and about 43% as attending church regularly.[3]

In my own university, the ratio of those studying the sciences versus the arts within the undergraduate population currently stands at close to 1.2:1. It has been an intriguing exercise to track the subject-areas studied by the student undergraduate population within my church in Cambridge over the past few years. The ratio of science to arts students has consistently been measured in the range between 2:1 and 4:1 — depending on the year — in other words scientists are found within this particular Christian community at double the expected frequency, or even more. Anecdotal evidence suggests that this disparity is quite common in other large city-centre churches attended by students. So there appears to be a selection pressure operating whereby either Christians are more likely to go into the sciences, or people studying science are more likely to become Christians, or both.

The NOMA Model

The *second model* that has been proposed to describe interactions between science and faith has been dubbed by Stephen Jay Gould the 'NOMA Model' that is, non-overlapping magisteria.

[3]   Stark & Jannaccone (1996). Various twentieth-century interactions between science and religion in America have been surveyed by J. Gilbert (1997).

To quote Gould (2002): 'The magisterium of science covers the empirical realm: what is the universe made of (fact) and why does it work this way (theory). The magisterium of religion extends over questions of ultimate meaning and moral value'. Gould therefore proposes a model in which science and faith effectively live in non over — lapping compartments, governed by their own particular questions and methods, with a water-tight bulkhead in between. The two enterprises are about quite different types of activity, and therefore there is no reason why they should come into conflict.

Now I have a great deal of sympathy with this view. There seems little doubt that science and faith *are* addressing rather different kinds of question — and we'll come back to a modified version of this Model in a moment. It fits rather nicely with the pragmatic departmental structure of our universities. Theologians let the scientists get on with their business and scientists let the theologians get on with their business and ne'er the twain shall meet. It makes for a quiet life — no bad thing.

But I am sure you will see that there are some serious problems with such a Model. For example, it is a matter of fact that there has been a constant traffic of ideas between science and faith during the emergence of the modern scientific movement. And of course Gould himself was well aware of that fact, writing some wonderful essays about historical figures in the history of science which underline this very point. A further problem with the NOMA model arises from the high proportion of religious believers within the scientific community. Most of us in that category do not wish to maintain our science and our faith separated by watertight bulkheads — even should such an exercise in mental gymnastics be physically possible — *rather we wish to integrate* our science and faith. The question is: can this be done and, if so, how?

The Integrationist Model

This brings us nicely to our *third model* to describe science-faith interactions, which we can call the Integrationist Model. This category contains a vast array of sub-models, but all can be subsumed under the 'Integrationist' heading, because they all have this in common: they integrate or fuse scientific or religious knowledge to certain degrees so that attempts are made to build religious systems of belief based on scientific knowledge or, conversely, scientific knowledge based on religious belief.

For example, my old tutor in biochemistry from Oxford days, Arthur Peacocke, in his recent book *Paths from Science Towards God*, argues that a rational set of theological beliefs can be *built upon* scientific knowledge and scientific ways of thinking (Peacocke, 2001; reviewed in Alexander, 2003).

In a rather different kind of example, Edward Wilson in his book *Consilience* envisages a day when all forms of knowledge will be united, but the strategy he chooses to adopt to bring about this fusion of all knowledge is that of scientific naturalism — religious beliefs in this view are simply transformed into sociobiology (Wilson, 1999). Whether such a stratagem can be justified is a point to which we shall return, for it lies at the heart of the question as to whether human values, and indeed ethics, can be extracted from scientific knowledge.

Just to illustrate how very diverse are the bed-fellows in the integrationist category, creationists also can to some extent be subsumed within the Integrationist Model, except that in this case the integration works in the *opposite* direction — they start with their particular interpretation of the Genesis text, a religious belief, and then use that to propose a rival theory of origins which they then present as a scientific theory, so-called scientific creationism.

The Complementarity Model

The *fourth model* to describe the interactions between science and faith, we can dub the Complementarity Model. This represents the view that science and religion are really trying to address rather different kinds of question which are not rivals to each other, but instead are complementary.

The model may be illustrated by the way in which various kinds of scientist study the human body, each using a different set of techniques and approaches. In principle, at the most basic level, a physicist could describe the body using the techniques and language of physics. But a description of the body at this level would be unbelievably complex — a description at the atomic level of the swirl of elementary particles and their energy relationships which comprise the ultimate building blocks of all living matter. As it happens the physics level of description would be quite inadequate to understand important biological concepts like 'cell' or 'DNA' — it's not that cells and DNA are not composed of atoms, its just that you need a higher, emergent level of description in order for such concepts to become meaningful. Such descriptions are provided by the biochemist and then the cell biologist. Then one level 'up' from cell biology comes physiology, the study of the dynamic interactions between the various organs of the body. For brainy primates like humans there is then the need for yet a further level of explanation which requires the language and techniques of psychology. After this level of explanation we would then wish to study the organism in the context of its interaction with the environment using the language and concepts of environmental biology and social anthropology.

Let's imagine now that we are being investigated by a bunch of super-scientists from Mars who have provided us with sets of absolutely complete descriptions of the human body at all these various levels that we have been considering. Would there still be further levels of explanation that our super-scientists would still like to explore? I think there certainly would be. For example, what does morality mean and how do humans justify their ethical decisions? How do they know what they *ought* to do? And in an ultimate sense why are there conscious beings anyway on planet earth? Why is there a universe with the particular properties that has made this strange phenomenon possible? These are the religious or philosophical kinds of questions. The answers to such questions are not in any kind of *rivalry* with the other levels of description that we have been considering — they are complementary to them. In fact, I would suggest, we all have to live, irrespective of our beliefs, as if we already knew the answers to such questions. In reality, we all have to live committed lives — it is a basic property of human existence.

The map-making analogy has been a useful one in thinking about the relationship between scientific and religious knowledge. If you want to describe a country adequately then you need separate maps for geology, communications, rainfall and so forth, an almost endless potential for representing the same geographical area from some new and specialised angle. It would be confusing to try and pack all the information on to a single map. Instead we make separate maps. The various maps are not rivals but provide complementary information. They are all about the same reality but viewed from different angles, just as the various levels of description of a human being are all necessary to provide a complete picture.

It was the neuroscientist the late Donald MacKay who popularised the Complementarity Model. MacKay asked the question: 'When are we justified in insisting that two pictures must be complementary?' To which he gave the answer: 'Only when we find both are necessary to do justice to experience' (MacKay, 1988, p. 35). So Complementarity is not some catch-all Model which is invariably appropriate in explaining the relationship between scientific and religious forms of knowledge, but it has been and continues to be a useful way of thinking about science-faith interactions. And what distinguishes the Model most from NOMA is that it insists that the explanations on offer are all *necessary* explanations to do full justice to the same reality. Biological knowledge alone really *is* insufficient to fully comprehend that highly complex entity — a human-being.

Two Views of Human Values

Having briefly reviewed these four types of Model to describe science-faith interactions, let's now go back over them and ask the question — in each case, how does the Model in question relate to human values?

The 'Conflict Model' certainly looks somewhat barren ground in this context. One problem is that the hyper-reductionist views that tend to be popular amongst people who adopt a conflict stance undermine rather than support concepts of human worth and dignity. If human beings are continually portrayed as nothing but naked apes, then perhaps one should not be overly surprised if people play their allotted roles accordingly. James Rachels was, I think, correct, when he wrote in his book *Created from Animals: The Moral Implications of Darwinism* (1990):

> The abandonment of lofty conceptions of human nature, and grandiose ideas about the place of humans in the scheme of things, inevitably diminishes our moral status. God and nature are powerful allies; losing them does mean losing something.

Thornhill and Palmer's book, *A Natural History of Rape: Biological Bases of Sexual Coercion* (2000), maintains that human males are by nature rapists, murderers, warriors and perpetrators of genocide. Likewise Ghiglieri in *The Dark Side of Man: Tracing the Origins of Male Violence* (1999) argues that rape is an adaptation to increase the reproductive success of men who would otherwise have little sexual access to women. The problem with such books is not that biology is irrelevant to such issues, but rather that they attempt to reduce complex multi-factorial social and moral issues down to fit a procrustean bed composed of biological knowledge alone. But biological knowledge alone is not up to such Herculean tasks. A common failure of scientists is to think that their narrow discipline is the Answer with a capital 'A' to a particular complex social or moral problem. Unfortunately this is not the case, and their naivety in making the attempt is embarrassing to other members of the research community.

If the Conflict Model, with its tendency to favour arch-reductionism, provides infertile ground for the generation of human values, then does the NOMA Model fare any better? This, you will remember, is the idea that science and faith should each go about their own independent business. Certainly, when compared with the Conflict Model it comes as a breath of fresh air. Theologians and moral philosophers may pursue their enterprises independently of science and scientists will value and utilise their insights. As a matter of fact, the great majority of scientists, I think, realise that scientific knowledge *per se* has little to say about moral and ethical issues nor, for that matter, about the worth and dignity of humans in relation to other animals. Typical of this view are the comments made by the London geneticist Steve Jones in his Reith Lectures a few years ago:

> Science cannot answer the question that philosophers — or children — ask: why are we here, what is the point of being alive, how ought we to behave? Genetics has almost nothing to say about what makes us more than just machines driven by biology, about what makes us human. These questions may be interesting, but scientists are no more qualified to comment on them than is anyone else (Jones, 1994, p. xi).

Whilst profoundly agreeing with such sentiments, the problem with the NOMA model is, I think, that in reality the separation between facts and values can never be quite as absolute as the Model suggests. Some scientific research areas are ethically more loaded than others. And again it is the same individual scientist who may both produce the new scientific knowledge that raises acute new ethical issues *and* who personally has to struggle with the very issues that are raised. A watertight separation of ideas is hardly a practical option under such circumstances. Furthermore, whereas biology or other sciences cannot, I think, dictate ethical decisions, accurate scientific information often has a *highly* relevant bearing on the moral decision-making process, so to tear the science too far apart from the theology and from the moral philosophy is a mistake.

An Integrationist Model — Michael Ruse

What about Integrationist Models? There is one Integrationist sub-Model to which we need to give particular attention in the context of human values. This is the idea proposed by evolutionary naturalism that an understanding of our evolutionary history does indeed generate a normative ethics, that is, an ethics that is morally binding for the whole of humankind. Such a proposal has been made by philosophers such as Michael Ruse in his book *Taking Darwin Seriously* (1986), and we need to spend some time in considering such a claim.[4]

Ruse tries to argue his way from biology to ethics in 5 steps: In 'Step 1' he maintains that complex human behaviours, such as moral decision-making processes, can be inherited. In 'Step 2', it is claimed that these innate dispositions have, or once had, adaptive value: they increased the chance of parents passing on their genes to their descendants. In 'Step 3' Ruse proposes that the force of the 'ought' which is implicit in all genuine ethical discourse is based on such innate biological drives derived from our genetic inheritance: 'Morality', says Ruse, 'is a collective illusion foisted upon us by our genes ... the illusion lies not in the morality itself, but in its sense of objectivity'. In 'Step 4' we are informed that such biological drives result in ethical impulses which, as a matter of fact, are broadly in line with traditional morality, promoting the 'values cherished by decent people of all nations'. Finally 'Step 5' of the argument tells us that we have a moral duty to aid the process of evolution since it has generated moral beliefs rooted in 'the very essence of living beings' which are truly international in scope.

---

[4]   The critique that follows was originally published in Alexander (1999).

A thorough critique of such a position is beyond the scope of this paper, although I have given the arguments some detailed attention in *Rebuilding the Matrix* (Alexander, 2001, see ch. 11). Suffice it to say for the moment that two types of critique may be levelled at Ruse's position, empirical and philosophical. Three of the steps (Steps 1, 2 and 4) make empirical claims. Steps 1 and 2 are not impossible in principle, but suffer from poor experimental support. There are currently no firm data supporting the genetic inheritance in humans of any complex forms of behaviour, certainly not of the inheritance of moral beliefs as Ruse's theory suggests. Those interested in this field may wish to read the recently published report from the Nuffield Council on Bioethics (2002) entitled *Genetics and Human Behaviour*. If there is no genetic basis for human moral convictions then neither, of course, can they have any inheritable adaptive value. An alternative position to that of Ruse agrees that moral convictions could have adaptive value, but via the fast process of cultural transmission, rather than by the 'slow-track' of genetic change. A commonly held view amongst biologists is that genetically encoded behavioural programmes are dominant amongst animals, but that in humans the acquisition of language and of conscious intellectual processes has enabled such a rapid transmission of behavioural norms as to make arguments based on slow genetic changes redundant.

In 'Step 3' Ruse starts to run into philosophical problems. Step 3, you will remember, is the proposal that the force of the 'ought' which is implicit in all genuine ethical discourse is based on innate, inherited biological drives. The 'is-ought' distinction (which Ruse refers to as the 'naturalistic fallacy'), pointed out forcefully by Hume and later expounded by the Cambridge philosopher G.E. Moore is not, I think, so readily circumvented. Moore pointed out that all attempts to justify moral claims by reference to descriptions of the physical world are doomed to failure. In short, you cannot derive an 'ought' from an 'is'. Ruse tries to side-step the fallacy by re-defining 'ought' so that the word no longer has its traditional sense of an implicit appeal to an objective yardstick of morality, but instead refers merely to innate dispositions. But in the process of re-defining 'ought' as a biological disposition, the force of the concept vaporises. Moral obligations founded on a disposition to do something are not really obligations at all. Furthermore, if our sense of objectivity about morality is, as Ruse claims, a genetically programmed illusion, now that this deception has been revealed by science, we can choose to ignore it.

The weakness of the naturalistic argument at this point is highlighted by Step 4 of Ruse's argument, since it is not clear how 'traditional morality' can be maintained by appeals to innate dispositions. In fact Step 4 — the claim that biological drives result in ethical impulses which are broadly in line with traditional morality — is empirically false. In practice people are not robots — they make genuine moral choices which vary from ethnic cleansing to caring for AIDS patients in the shanty-towns of Johannesburg. Moral systems of belief vary enormously in different cultures and at different historical periods. The *biological* perspective is simply that people have different urges to do different things, but biology provides no criteria for deciding why one set of urges should be labelled more 'moral' than another. Armed only with Ruse's presuppositions, we could be left describing the atrocities of the Nazi regime as yet another 'interesting' manifestation of humankind's innate dispositions. The value of the human individual is, I think, at severe risk within Ruse's biological framework.

I have spent some time on this issue because the recent flourishing of genetics — which I applaud — has at the same time aroused public anxieties about the implications and scope of this flood of new information. I think philosophers, biologists and others need to be very cautious about making wild extrapolations from biological data into the realms of human goals and values. The history of eugenics is not a pretty one.

Human Values —
The Complementarity Model

Lastly, how does the Complementarity Model fare when it comes to issues of science and human values? I think it fares rather well. We have already observed that science is quite restricted in the kind of knowledge that it generates. Science describes. It quantifies. It makes generalisations based on a large number of observations. It is highly successful at constructing a body of reliable knowledge describing the physical world and its properties. But at the same time its descriptions *ipso facto* leave out great swathes of human knowledge and endeavour — ethics, aesthetics, our own personal biographies, questions of ultimate value and meaning, the justification of religious beliefs and much more besides. In the complementarity model these are all vital spheres of human life and experience. To use Donald Mackay's criterion, yes we do need the explanatory powers of *all* these various aspects of human experience to do justice to the reality of our own existence as conscious observers of a remarkable universe.

The Complementarity Model therefore highlights the point that the sources of human values do not arise from scientific knowledge *per se*. Where, then, do they come from? There are others far more capable than I to map out the various sources of our moral beliefs. I note that Michael Fuller, Principal of the Theological Institute of the Scottish Episcopal Church, has provided a useful overall schema of the various positions in his article 'Science, Religion and Ethics' (2002). However, at this stage I would like to return to the comment made by James Rachels that I quoted earlier: 'God and nature are powerful allies; losing them does mean losing something'. What is it that we lose by abandoning belief in God as creator and as the ultimate source of human values?

When I was a student at Oxford in the mid-1960s the biggest society in the university was the Humanist Society. The majority view within the society was that Christian theism was outmoded. Nevertheless the majority opinion also maintained that the values based on human dignity which Christianity had underpinned over the centuries were worth keeping, and it was these that the Humanist Society intended to preserve. In the decades since the 1960s the Humanist Society has disappeared almost without trace and the extreme difficulty of building a rational basis for human dignity without an underpinning metaphysical framework has become increasingly apparent. If you start, for example, with a purely biological account of the individual human as the most important story that can be told about that person, then the rational maintenance of human value becomes problematic.

Infanticide — A Consequentialist and a Theistic View

A controversial debate about infanticide has recently focused attention on how rival metaphysical worldviews generate very different understandings of human value. The debate is useful because it sheds light on how a given presupposition can lead inexorably, and in a highly rational manner, to a conclusion that many people find deeply abhorrent. I choose this example not because there is time here to do justice to such an important discussion in itself, but because it provides a clue as to how similar public debates may well go in the future when it comes to further questions, such as whether to clone humans, or genetically modify the human germ-line, or generate human-monkey hybrids in order to carry out research on brain function and consciousness.

Peter Singer, an advocate of infanticide, is both accurate and explicit concerning the implications of rejecting a theistic framework for human value:

> If we go back to the origins of Western civilisation, to Greek or Roman times, we find that membership of Homo sapiens was not sufficient to guarantee that one's life would be protected.... Greeks and Romans killed deformed or weak infants by exposing them to the elements on a hilltop. Plato and Aristotle thought that the state should enforce the killing of deformed infants.... The change in Western attitudes to infanticide since Roman times is, like the doctrine of the sanctity of human life of which it is a part, a product of Christianity. Perhaps it is now possible to think about these issues without assuming the Christian moral framework that has, for so long, prevented any fundamental reassessment. (Singer, 1993, pp. 88, 173).

Singer is of course correct: Infanticide was so common in the Graeco-Roman world that one contemporary historian, Polybius, writing in the second century BC, concluded that it had contributed to the depopulation that had occurred in Greece at that time. Singer is also correct in his suggestion that infanticide declined due to the Christian insistence on the sanctity of life. Following the conversion of the Emperor Constantine to Christianity in AD 313, infant exposure was made punishable by law.

What, therefore, does Singer wish to put in place of the sanctity of life? Singer is a consequentialist — he believes that ethical decisions should be based on an assessment of their overall consequences. His fundamental ethical starting point is the 'principle of equal consideration of interests', the belief that the interests of all human beings must be taken into account when assessing the consequences of our actions. This principle extends also, says Singer, to sentient animals, those animals who can suffer or be 'happy'. Only a being that can suffer can be said to 'have interests'. Human beings can be considered in two quite distinct ways — either as belonging to the species *Homo sapiens*, or by being a person. A person is a 'self-conscious or rational being' who can therefore act as an agent in making decisions, capable of thwarted desires for the future. In addition Singer wishes to argue that primates, for example, are also self-conscious, at least to some extent, and should be included in the term 'person'. Therefore being a member of the species *Homo Sapiens* is neither necessary nor sufficient for being a 'person'.

What are the conclusions of such a starting point? It is not intrinsically wrong to kill a new-born baby because the baby is not yet self-conscious whereas it would be wrong to kill adult animals who are supposed to be self-conscious. 'So it seems', writes Singer,

> that killing, say, a chimpanzee is worse than the killing of a human being who, because of a congenital intellectual disability, is not and never can be a person ... the life of a new-born baby is of less value to it than the life of a pig, a dog, or a chimpanzee is to the non-human animal.

In this same passage Singer goes on to urge us that

> we should put aside feelings based on the small, helpless, and — sometimes — cute appearance of human infants.... [L]aboratory rats are 'innocent' in exactly the same sense as the human infant ... killing a disabled infant is not morally equivalent to killing a person. Very often it is not wrong at all.

What happens when we carry out a comparable assessment of the disabled new-born, but now with reference to theism? Why is Christian theism, for example, so hostile to infanticide? There are two main inter-connected reasons. First, each human life is seen as a gift from God towards which the human community has commitments and duties. These duties are particularly pressing when the individual is helpless and therefore entirely dependent on the human community for support. The love shown by the human community should reflect something of God's love for each individual, a love which operates independently of the person's physical status. The helpless new-born therefore has an intrinsic value which is independent of its biological status, reflecting God's love mediated by the commitment of the supportive human community.

Second, in Christian theism humans are made 'in the image of God' (Genesis 1.27). This expression is first introduced in the Biblical text in the context of the responsibilities given to humankind to care for the earth and its biological diversity. In this scenario, humans are delegated by God to be his 'earth-keepers'. The 'image of God' is not therefore so much a static concept, referring to human reason, or free − will, or other particular intrinsic human qualities, but rather to the dynamic relational status of humans to God, in particular regarding their delegated responsibilities. These moral responsibilities are given not just to a few individuals, but to the whole of humankind. A severely disabled infant, for example, may never be able to contribute very much, if anything, individually to fulfilling these delegated responsibilities, but nevertheless is part of the human community that as a whole carries this moral obligation upon its shoulders. Bearing the 'image of God' is thus about relationship and solidarity, once again pointing not to the genetic perfection or blue eyes of the helpless infant, but rather to its intrinsic worth as a member of this moral community. All humans are God's image-bearers in this sense and are our neighbour. Jesus says that we are to love our neighbour as ourselves.

I do not intend by making such a contrast to suggest that all consequentialist arguments are wrong and that all moral decisions informed by Christian ethics are black and white − far from it. Consequentialist arguments can be of real use in trying to decide between the lesser of two evils. And when it comes to the nitty-gritty issues, Christians sometimes disagree. But − the metaphysical framework within which ethical reflection is carried out really does make a difference. Over the coming decades we will be facing numerous novel ethical dilemmas, many of them thrown at us by the fast pace of biological research. Science, I would suggest, will only maintain a 'human face' if there is a metaphysical system in place robust enough to provide a matrix within which intrinsic human value can be sustained. For myself, the Christian world-view provides such a framework, one which is both intellectually satisfying and ethically coherent.

Dedication

This lecture [given in the University of St. Andrews on 15th November 2002] is dedicated in memory of my two elder brothers, John and David, who died on 2nd and 13th November, 2003, respectively, shortly before this lecture was given. As co-founder of Lion Publishing, David in particular was instrumental in stimulating my interest in writing on the subject of science and faith.

References

Alexander, D.R. (1999), 'Can science explain everything? Scientific naturalism and the death of science', *Cambridge Papers* Vol 8, No. 2, June.

Alexander, D.R. (2001), *Rebuilding the Matrix: Science & Faith in the 21st Century*. Oxford: Lion.

Alexander, D.R. (2003), *Science & Christian Belief*, **15**, pp. 198–200.

Alexander, D.R. & White, R. (2004), *Beyond Belief : Science, Faith and Ethical Challenges*. Oxford: Lion.

Atkins, P. (1992), 'Will science ever fail?' *New Scientist*, August, pp. 32–35.

Berger, Peter L. (1996), 'Secularism in Retreat', *The National Interest*, **46** (3), Winter 1996/97.

Brooke, J.H. (1991), *Science & Religion : Some Historical Perspectives*. Cambridge: CUP.

Brooke, J. &. Cantor, G. (1998), *Reconstructing Nature : The Engagement of Science and Religion*. Edinburgh: T & T Clark.

Dawkins, R. (1991), 'The Ultraviolet Garden', Royal Institution Christmas Lecture No 4.

Fuller, Michael, (2002), 'Science, religion and ethics', *Expository Times*, November.

Gilbert, J. (1997), *Redeeming Culture: American Religion in an Age of Science* University of Chicago Press.

Gould, S.J. (2002), *Rocks of Ages: Science and Religion in the Fulness of Life* Ballantine Books.

Harrison, P. (1998), *The Bible, Protestantism and the Rise of Natural Science* Cambridge University Press.

Hooykaas, R. (1972), *Religion and the Rise of Modern Science*. Edinburgh: Scottish Academic Press.

Jones, S. (1994), *The Language of the Genes*. Flamingo.

Larson, E.J. & Witham, L. (1997), 'Scientists are still keeping the faith', *Nature*, **386**, pp. 435–436.

Larson, E.J. & Witham, L. (1998), 'Leading scientists still reject God', *Nature*, **394**, p. 313.

236   *Science, Consciousness & Ultimate Reality*

Lindberg, D.C. & Numbers, R.L. ed. (1986), *God & Nature: Historical Essays on the Encounter Between Christianity and Science*. University of California Press.

MacKay, D.M. (1988), *The Open Mind*. Leicester: IVP.

MacKenzie Brown, C. (2003), 'The conflict between religion and science in light of the patterns of religious belief among scientists', *Zygon*, **38**, pp. 603–632.

Nuffield Council on Bioethics (2002), *Genetics and Human Behaviour: The Ethical Context*. Obtainable for £3.00 by emailing: bioethics@nuffieldfoundation.org

Peacocke, A.R. (2001), *Paths from Science Towards God: the end of all our Exploring*. Oxford: Oneworld.

Rose, S. (1988), 'Reflections on reductionism', *Trends in Biochemical Sciences*, **13**, pp. 160–162.

Ruse, M. (1986), *Taking Darwin Seriously*. Oxford: Basil Blackwell.

Russell, C.A. (1985), *Cross-Currents, Interactions Between Science and Faith* Leicester: IVP.

Russell, C.A. (1989), 'The Conflict Metaphor and its Social Origins', *Science & Christian Belief*, **1**, pp. 3–26.

Singer, P. (1993), *Practical Ethics*, 2nd edn. Cambridge University Press.

Shapin, S. (1996), *The Scientific Revolution*. University of Chicago Press.

Stark, R. and Jannaccone, L. (1996), *American Economic Review: Papers and Proceedings*, p. 436.

Thornhill, R. & Palmer, C. (2000), *A Natural History of Rape: Biological Bases of Sexual Coercion*. Boston, MA: MIT Press.

Wilson, E.O. (1999), *Consilience: the Unity of Knowledge*. Abacus.

John Habgood

Science and Human Responsibility

On Being a Person

Introduction —
The Possible and the Permissible

The only other occasion when I have had the pleasure of speaking after Lord Winston was a brief exchange in the House of Lords a few years ago. This was about the case of Diane Blood, the woman who wanted to use her dead husband's sperm to conceive their child. She eventually achieved this in the Netherlands, and is now pregnant with his sperm for a second time. Lord Winston generously undertook to introduce a Bill on her behalf, which would have made post-mortem conception legal in this country. I, among many others, resisted the idea, and I did so on the grounds that such legislation would entail a major change in the legal boundaries within which human life has hitherto been lived — the final boundary being death.

Doctors rightly pay great attention to the needs of individual patients, and struggle to find more and more ingenious ways of helping them. This is their vocation. Legislators, on the other hand, have to ask themselves, what are likely to be the long-term social consequences of new techniques as these become generally available? Does the fact that something is possible mean that it should also be permissible? And such questions too are surely right. From an ethical or religious perspective one might want to ask a rather different kind of question. What might the wide employment of this or that technique do to our understanding of ourselves and each other as persons? Is it compatible with respect for human dignity, and with what we know of God's purposes for human beings? It is from these broader, more philosophical and religious perspectives that worries can arise about attempts to circumvent the finality of death.

At first sight such concerns might seem a rather abstract. But imagine a future social order in which sperm and eggs were regularly preserved and available for the conception of children from genetic parents who had long been dead. Imagine the competition for genes which were felt for some reason to be specially valuable. Or try to enter into the minds of donors who wished in some way to preserve their ability to reproduce themselves far into the future. Whatever the attraction of such possibilities to those for whom individual choice takes precedence over everything else, the existence of a kind of car boot sale of desirable genetic material, could not fail to change attitudes towards the children destined to be its eventual products. The word 'commodities' is frequently used — and with reason. Even the much more modest proposal to confine post-mortem conception to dead husbands could subtly complicate the relationship between parents and children, making the children into orphans by design, rather than orphans by accident.

Leaving aside the problems concerning dead donors, there is, of course, already a flourishing trade in living ones. What are we to make of the web site, for instance, on which it is possible to bid in an auction for eggs and sperm produced by glamour models? The American owner of the site reckons that eggs will sell for 15,000 to 150,000 dollars and sperm for 10,000 to 50,000 dollars. What might have been the price, I wonder, if Einstein, say, or Marilyn Monroe, had conveniently left a donation? When pressed to justify his trade the owner of the site, a fashion photographer, replied, 'We bid for everything else in this society — why not eggs?'

All this might be dismissed as nonsense on the lunatic fringe of a corruptingly affluent society. But there is an echo of it in the questions currently being asked about whether the donors of sperm for artificial insemination should be identifiable. Those conceived by such means have an obvious interest in knowing where they have come from, and can argue that if adopted children are allowed to discover their parentage, so should they. But the parallel between adoption and artificial insemination is not exact. In adoption there is no difficulty in distinguishing between genetic parents and social parents, and any emotional complications should be explicit from the beginning, at least to the parents. But in artificial insemination the distinction is much more difficult. The donor is an invisible third party within the parents' marriage, as well as almost certainly being a third party in many other marriages or partnerships, and the father of many other children. There is a real conflict of interests in current attempts to make the donor identifiable. What might the implications be, I wonder, for marriages in which this new partner occupies a much more intimate place, between husband and wife, than is the case with the parents of an adopted child? And the children themselves might find it hard to handle that they were part of an extensive network of half-siblings, grandparents and others, with whom their genetic relationship was just as strong as with those whom they have regarded as their own family.

I raise these issues, not because I have any definitive answers, but as current illustrations of the way in which new techniques, devised with the best of intentions, can have social repercussions which touch fundamental aspects of what it is to be a person. There are even more striking examples if one considers the future possibilities of genetic manipulation, but these would take us too far afield, and there is already a growth industry in books about it.

Tony Bland — Persistent Vegetative States

A few years ago, when I wrote a book with the title *Being a Person*, I was stimulated to do so by the problem of knowing when a person really is dead. My concern centred on the much publicised case of a boy, Tony Bland, who was in a Persistent Vegetative State as a result of being suffocated during the 1989 disaster at the Hillsborough football ground. He was unconscious, in fact brain dead, but he could breathe without assistance, and his body, though immobile, looked quite normal. He had been in that state for more than two years, during which he was visited every day by his parents. He could have survived in that state for many more years, but eventually they asked that what remained of him should be put to death, a wish which after much legal wrangling was finally granted.

In Tony Bland's case we have the opposite problem to that of post-mortem sperm. Frozen sperm, eggs, and embryos can claim to cheat death by preserving people's reproductive capacity apart from their bodily presence. PVS describes a condition in which the living body is preserved when the conscious life of the person has been destroyed. Both conditions defy the normal limits of human life and personality. They maintain a quasi-personal presence when the wholeness of personhood, as expressed in what one might call 'embodied consciousness', has been irretrievably broken. Both pose urgent questions about what we actually mean by 'persons', and about how this understanding should inform our decisions in the increasingly complex areas of reproductive and medical ethics.

In the case of the young PVS patient I was struck by the extent to which some residual degree of personhood was maintained in being by the continuous attention of his parents. They were responding to a living body which was still that of their son, and which needed their loving care. He was, as it were, still held as a person in bodily relationship with them, just as he would have been had he been able to respond. It was this close relationship which gave meaning to what was left of him, even though he was totally unresponsive. Thus when his parents decided that it was a relationship with no future, something died, and they wanted to see that death fully implemented.

We all live in, and are formed by, our relationships with others and this is in part what is meant by being a person. Such relationships can, I believe, and do, transcend death, but they cannot healthily continue with what is effectively a dead body — otherwise we might find ourselves in the world of Hitchcock's Psycho. This is why the rituals surrounding death are so carefully designed to allow a transference from one kind of relationship to another. We love those who have died, but need to let them go. We treat dead bodies reverently and tenderly, but then we dispose of them and, if we are believers, we have to learn to know the persons that they once were, as being held in the love of God, who still has a future for them.

Consciousness and Relationships

What worried me about Mrs Blood and her dead husband's sperm, is that she was not letting him go. She was using him, even to the point of taking his sperm posthumously — not a pretty procedure. Maybe she is spiritually mature enough to cope with the contradictions in this kind of relationship, and I am happy for her in her present happiness. To recognise in your child some of the characteristics of the person you have loved is a great blessing. But to prioritise genes over the reality of a living relationship seems to me too near for comfort to the kind of egg and sperm market I mentioned earlier. Genes are important, but they are not everything. Our total genetic identity may endow us with some of our physical characteristics, and with some traits and predispositions rather than others, but what we actually become depends on what happens to us, what we do, the cultural and physical environment in which we develop, and the personal relationships within which we stand. This is why in most instances what happens to us in life is more significant than our precise genetic inheritance, and why some people's anxiety to know their true parentage, or to buy superior genes for their children, may have been exacerbated by the current prominence given to all things genetic.

I have chosen to approach the question of being a person in this way, partly because of my previous modest encounter with Lord Winston, and partly because at one stage I had much to do with PVS, and was strongly influenced by it in my thinking about personhood. I have been concerned to make the point that if we are to safeguard what is precious about human personality, then we have to take the whole of what we are into account: genes, bodies, relationships, culture, personal history, environment in the widest sense, and ultimately — as I shall argue — our relationship with God. Human personality is the point of intersection, as I see it, where both science and religion are relevant as we try to make sense of the mysterious reality of being a person, and of our ability to recognise and respond to the personhood of others.

At the focal point of this intersection between science and religion, there is the mystery of consciousness. Somehow our extraordinary process of development generates an inwardness, a subjectivity, which is unique to each of us. We can recognise it in others, and we can mark its absence, as in the case of Tony Bland. What we cannot do is to observe it, except in our own direct experience. In trying to observe it and study it we inevitably objectivise it, and thus destroy its character as irreducibly subjective. We can, of course, study thinking and feeling; this is what psychologists do. We can also study brain mechanisms, whether at the level of neurophysiology, or through the cognitive sciences, or in more general terms by trying to trace the influence of our evolutionary past. We can even trace the activity of different parts of the brain by observing changes in the blood supply, as it engages in different mental processes. What we cannot do is to know exactly what another person is thinking or feeling, we cannot share their consciousness of it, other than through interpreted experience.

This is not just an oddity. The distinction between subject and object, personal awareness and public knowledge, is fundamental to the whole process of knowing. It need not imply the kind of dualism for which Descartes was responsible. Thinking, as I have already said, is not beyond the reach of objective study, even less so now that information technology has made it possible to create machines with a modicum of intelligence. A chess-playing computer may reach its conclusions by a much more laborious process than the human brain, but it is hard to deny that some equivalent to thinking is taking place.

Nobody is supposing, though, that such a computer has a subjective point of view. It is clear also from their behaviour that many animals think, though what they experience in terms of self-awareness cannot be known by us. We human beings have developed instruments, notably language, which enable us to communicate some of the inwardness of direct personal consciousness. Our personal formation and the growing awareness of selfhood depends on all the genetic, social, cultural and environmental factors I have been concerned to stress, in trying to spell out what makes persons special. But consciousness itself remains the great impenetrable mystery, because it is not possible in principle to grasp its essential character objectively. The difficulty is that it becomes something else as soon as it is studied from the outside.

But this does not mean that nothing can be said about it. The series of consultations, of which today's meeting is a part, is intended to illustrate just how much the serious study of consciousness can tell us, both about ourselves as persons, and about the nature of ultimate reality. One possible approach to this larger agenda is by speculating how subjective awareness might have evolved.

Language must have had a part to play, especially the use of language, not just to convey formation, but to create a rapport between different speakers. One of the things we do by talking is to mark out for each other areas of shared experience and feeling. If I say, 'It's lunchtime', I am conveying information. But if I say 'That was a good lunch!' to someone who has just eaten with me, I am probably not telling them anything they don't know already. I am using language expressively to share common experiences and feelings. I am showing something of myself in the hope that it will find a response. It is through this kind of interchange, trivial in one sense, but immensely important in another, that social bonds are formed and cemented. Hence the importance of gossip. And it is no coincidence that women are usually better at it than men, because women tend to be the main formers of community.

From an evolutionary perspective it is not hard to see how this kind of self-expression in social animals can be just as necessary to survival as the ability to convey information about where food is to be found, or dangers avoided. It does not necessarily have to be in words. Gestures, facial expressions, mutual grooming and so forth, were probably the preliminary building blocks of social existence. Language enormously expanded the possibilities of social interaction. It is in such interchanges with others that personal formation takes place; people — as we say — discover themselves, and come to believe that others are conscious agents, just as they themselves are.

Language can also provide an interpretation of the inwardness of others, which in turn helps us to be more deeply conscious of our own.

This is presumably why so many young people today are glued to their mobile phones, saying nothing of any significance, but simply expressing their thoughts and feelings to one another, as part of the process of growing up.

Another hugely important gift of language is its capacity to take us out of ourselves, to escape from the here and now, to imagine what has been or might be, and to locate our own subjectivity within the larger context of a public world. In fact it is a vital instrument in enabling the distinction to be made between our immediate experience and the shared public world. I am fond of Edwin Muir's poem, *The Animals*, in which he imagines a world without language, on the fifth day of creation.

> *They do not live in the world,*
> Are not in time and space.
> From birth to death hurled
> No word do they have, not one
> To plant a foot upon,
> Were never in any place.
> For with names the world was called
> Out of the empty air,
> With names was built and walled,
> Line and circle and square,
> Dust and emerald;
> Snatched from deceiving death
> By the articulate breath.
> But these have never trod
> Twice the familiar track,
> Never never turned back
> Into the memoried day.
> All is new and near
> In the unchanging Here
> Of the fifth great day of God,
> That shall remain the same,
> Never shall pass away.
> On the sixth day we came.

Not very good science. But I think he was right about the power of language to shape our world, and to develop in us a sense of time, and change, and continuity. In demonstrating something of the character of our minds, it does not take us far, though, in trying to understand how our inner subjective world might have evolved. The ability to communicate through symbols seems already to presuppose a degree of subjectivity. I believe the clue lies in the nature of the environment in which evolution takes place.

Environment, Language and Evolution

Everybody knows about the importance of random variation and competition. But the crucial role of the environment, and of environmental change, often seems to take second place in people's understanding of how we have come to be. By environment I mean not just the natural world, but all that environs an organism, all that impinges on its life and affects its chances of passing on something to it successors. A social environment, for instance, may spell the difference between success and failure, which is why the social interactions I have been referring to, and the social structure to which they give rise, may be just as important in evolutionary terms as the quality of food in this or that part of the forest.

Most animals occupy environmental niches in which their distinctive characteristics give them an advantage over others, and some have learnt to modify their niches to give themselves still greater advantage. Birds which build nests rather than just deposit their eggs on the ground are an obvious example, and beavers' dams an even more striking one. There is no need, therefore, to think of environment as a mere backcloth. There can be active engagement with it, decisively so in the case of human beings. In fact one of our most distinctive human characteristics is our ability to adapt almost any environment to suit our needs. Nor is this just a matter of physical needs. We transform our surroundings to create aesthetic satisfaction, for instance, or to arouse excitement, or to heighten the quality of our experience. In short, we are uniquely interactive, not only with each other, but also with our environment, as understood in the widest possible sense. Furthermore the feedback between ourselves and a partially self-created environment enormously enhances its significance in making us what we are.

What has this to do with our subjectivity? Quite a lot, I believe, if we ask more searchingly where the notion of environment ends. What is it to which we ultimately relate? Does it make sense to imagine some kind of all-encompassing environment, by interaction with which human personality has been stimulated to develop, and in relation to which our conscious life is now lived? What I am suggesting is that the nature of the ultimate reality which environs us must at one level or another have a decisive role in shaping what we are. Thus if our subjectivity is an irreducible aspect of being human, it seems to follow that whatever we mean by ultimate reality must somehow include the possibility of subjectivity, and provide the scope and stimulus for its evolution.

I know what you are thinking — he's going to bring in God. It would indeed be easy to do so, and to point to him as one whose presence is an abiding feature of the reality within which our humanity has been shaped. But I am going to disappoint you. It would be too easy. Appeal to God in this kind of exploration is the equivalent of a short circuit. I am still concerned with what, if anything, an ordinary scientific approach might be able to tell us. It cannot handle subjectivity itself, as I have already tried to explain. But it might be able to tell us something about environments in which a kind of subjectivity could develop.

I return to what I was saying earlier about language, its power to open up other worlds and other minds, and I return also to the seemingly endless capacity we possess for creating and modifying our own environments. The odd thing about our creativity is that, like language, it extends well beyond the obvious necessities for survival. Art is an example, but so is play, and play is perhaps the more interesting because we can see its beginnings in the non-human world. It may have its roots in the need to hone physical skills, but even in animals some forms of play appear just to be fun. It is an expression of freedom, and escape from necessity, one of the many signs that life, especially human life, can burst its strict biological boundaries. There are constant reminders, too, of our human capacity and yearning for what transcends us, the impulse to rise above our circumstances, which is strangely at odds with the iron rule of law which is supposed to reign throughout physical reality. How can such things be, unless there is somewhere an openness built into the way things are, a possibility of freedom and creativeness, a space within which it is possible to encounter whatever it is that transcends our immediate bodily experience?

Createdness

A useful starting point, in my view, is to introduce the concept of createdness. Existence is not self-explanatory. Createdness implies that things exist by virtue of a reality which transcends them — a claim which may ring scientific alarm bells. But the kind of createdness I have in mind allows for continuing rational investigation without causing an instant theological short-circuit. One of its characteristics is a certain distancing between the created world and God. God, as it were, lets creation be itself.

Indeed in some respects he lets it make itself. Evolution, for instance, has frequently been interpreted by Christian theologians in this way. A recent symposium called *The Work of Love*, edited by John Polkinghorne, elaborates the point by drawing heavily on the idea of Christ's self-emptying in the incarnation. Here we can see God accepting limitations, disclaiming all power but the power of love. Though sustaining everything in being, he is not the direct cause of everything that happens. There is a degree of autonomous freedom within creation, and it is this which gives it meaning and purpose. Its actuality could not have been deduced from general principles. History makes a difference. The story of creation is not just the outworking of some preordained plan. To understand it requires empirical study; hence the importance of science.

But though God distances himself from creation, he is not absent from it. In Christian belief his presence and power are available at every point, as a gracious gift for those who seek it. And though he transcends creation, he can be known in and through created things, a possibility on which the whole of incarnational and sacramental Christianity depends.

Let me make clear what I am doing. Though for the last few minutes I have been talking in theological terms, I am not trying to prove the truth of Christianity by these means. I am asking whether this way of understanding the nature of ultimate reality can throw any light on our experience of subjectivity. I believe it can. I do not think it is capable of proof, but I put it forward as a better way of making some sense of subjectivity than any currently on offer.

Createdness, as I have said, entails a relationship with God as the one who sustains everything in being, who does not intrude upon its freedom to be itself, but can be known by those who seek him. It is a relationship which is mirrored at a less profound level in our existence as social beings. In order to become persons we are dependent on other persons, not only for our very existence, but for the countless personal interactions which enable our personhood to develop. There is an element of freedom within these interactions, as well as dependence. But within human relationships there is always the lurking fear that what we think of as 'ourselves' could be no more that the composite image of all these external influences. Hence the desire to be different, to transcend our circumstances, to assert a stronger sense of the self. Why? Because such a possibility of transcendence is in fact open to us, not only in our relationships with other people — as in the experience of loving and being loved — but is inherent in the possibility of relationship with the very source of our being — God. I say 'possibility' though, of course, we are related already by virtue of our createdness; but perhaps it is characteristic of this most basic kind of dependence on God that it, like the createdness of our environment, should open us to the desire for more. The point is that dependence on God is not a diminishment of us, nor the basis of a false self-image, as over dependence on other people's image of us might be. It is a recognition of what at the deepest level of our existence we really are. The Bible describes the experience of the presence of God in terms of 'knowing as we are known', and most actual religious experience echoes this.

Roots of Subjectivity

Perhaps it is here that we can begin to discern the roots of our subjectivity. I am suggesting that it belongs to the nature of created reality that it should bear the imprint of its transcendent origin. At the most basic level of all, the sheer givenness of things confronts us with the ultimate mystery of existence. At higher levels of complexity other qualities emerge, implying other dimensions of existence, which likewise owe their possibility to creation's openness towards and dependence on the One who transcends it.

In speaking of knowing and being known by God, I can't help seeing the great I AM passages in the Bible (I AM THAT I AM — Before Abraham was I AM) as insights into God as the Universal Subject of all experience, in whom knowing and being are one. I am aware of the danger of over-interpreting much disputed passages. But in modern human terms 'I am' is the basic expression of consciousness, an assertion that we are subjects not objects. What I am suggesting is that, in relationship to the great I AM of God — his universal subjecthood, both transcendent and incarnate — our own possibility of self-awareness is called into existence. Because the basis of all reality is the ultimate subject who knows us from within, we too can know ourselves as subjects.

It is obvious that in terms of our evolutionary history this self-awareness does not require explicit knowledge of, or explicit relationship with. God. How could it? Explicit knowledge and relationship are a long way down the road of mental development. But I return to what I was saying about the implications of our createdness, a mode of being dependent on the continuous sustaining will of God. Whether we are aware of it or not, such creaturely dependence is the most fundamental feature of the environment in which we live and move and have our being. On this understanding of what we are, it should not be too surprising that God as the source of being and knowing should draw out from creatures with capacities to receive them, the possibilities of freedom, self-transcendence, and self-awareness.

Perhaps this is one way of understanding what it means to be made in the image of God. But it may also be a reason for caution before throwing off too easily the limitations which this history has placed upon us. It is part of the glory of science, and of our huge technological capabilities, that we have been led to share some of the qualities of our Creator. This does not mean, however, that we have been given licence to cross all boundaries, even to the point of creating future generations in our own preferred image. Playing the creator to one another by exercising our powers to manipulate them, is surely the very opposite of being like the God revealed to us in Jesus Christ.

Journal of Consciousness Studies

www.imprint-academic.com

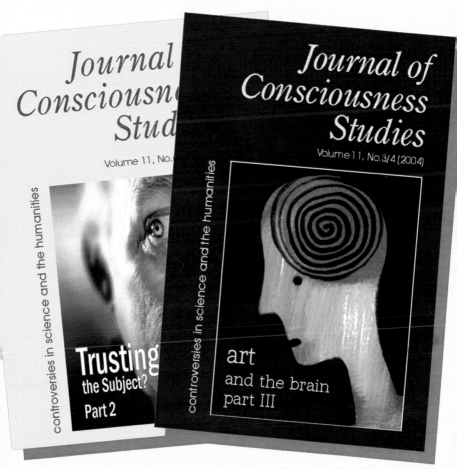

Journal
Consciousn
Stud

Volume 11, No,

controversies in science and the humanities

Trusting
the Subject?
Part 2

Journal of
Consciousness
Studies

Volume 11, No.3/4 (2004)

controversies in science and the humanities

art
and the brain
part III

'With *JCS*, consciousness studies has arrived'
Susan Greenfield, *Times Higher Education Supplement*